P9-DNE-743

Voyages of Discovery

Published by Firefly Books Ltd. 2008

Created by Co & Bear Productions (UK) Ltd.

First printing

Publisher Cataloging-in-Publication Data (U.S.)
Rice, A. L.
Voyages of discovery : a visual celebration of ten of the greatest natural history expeditions / Tony Rice ; introduction by David Bellamy.
[336] p. : col. ill., col. photos., maps ; cm.
Includes bibliographical references and index.
Summary: Paintings and photographs from the last three centuries document the record of ten of the most significant natural history expeditions. The stories behind these images — of explorers, naturalists, artists and photographers — entwine into a study of human achievement and natural wonder.
ISBN-13: 978-1-55407-414-3
ISBN-10: 1-55407-414-2
1. Natural history — History. 2. Voyages around the world. I. Bellamy, David. II. Title.
508/.09 dc22 QH15.R44 2008

Library and Archives Canada Cataloguing in Publication
Rice, A. L.
Voyages of discovery : a visual celebration of ten of the greatest natural history expeditions / Tony Rice ; introduction by David Bellamy.
Includes bibliographical references and index.
ISBN-13: 978-1-55407-414-3
ISBN-10: 1-55407-414-2
1. Natural history — History. 2. Voyages around the world. I. Title.
QH15.R44 2008 508.09 C2008-900770-0

Published in the United States by
Firefly Books (U.S.) Inc.
P.O. Box 1338, Ellicott Station
Buffalo, New York 14205

Published in Canada by
Firefly Books Ltd.
66 Leek Crescent
Richmond Hill, Ontario L4B 1H1

Front cover: Purple-faced leaf monkey (*Semnopithecus vetulus*), painted by Pieter de Bevere.
Back cover: The First Fleet entering Rio Harbour, painted by George Raper.
Spine: Collared scops owl (*Otus bakkamoena*), painted by Pieter de Bevere.
Title page: Specimen from one of Sir Hans Sloane's famed Herbarium volumes.

Printed in China

Voyages of Discovery

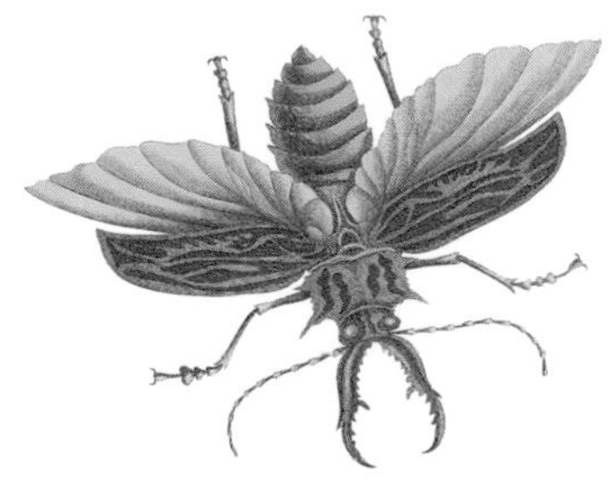

A Visual Celebration of Ten of the Greatest Natural History Expeditions

Dr. Tony Rice

Introduction by Dr. David Bellamy

Firefly Books

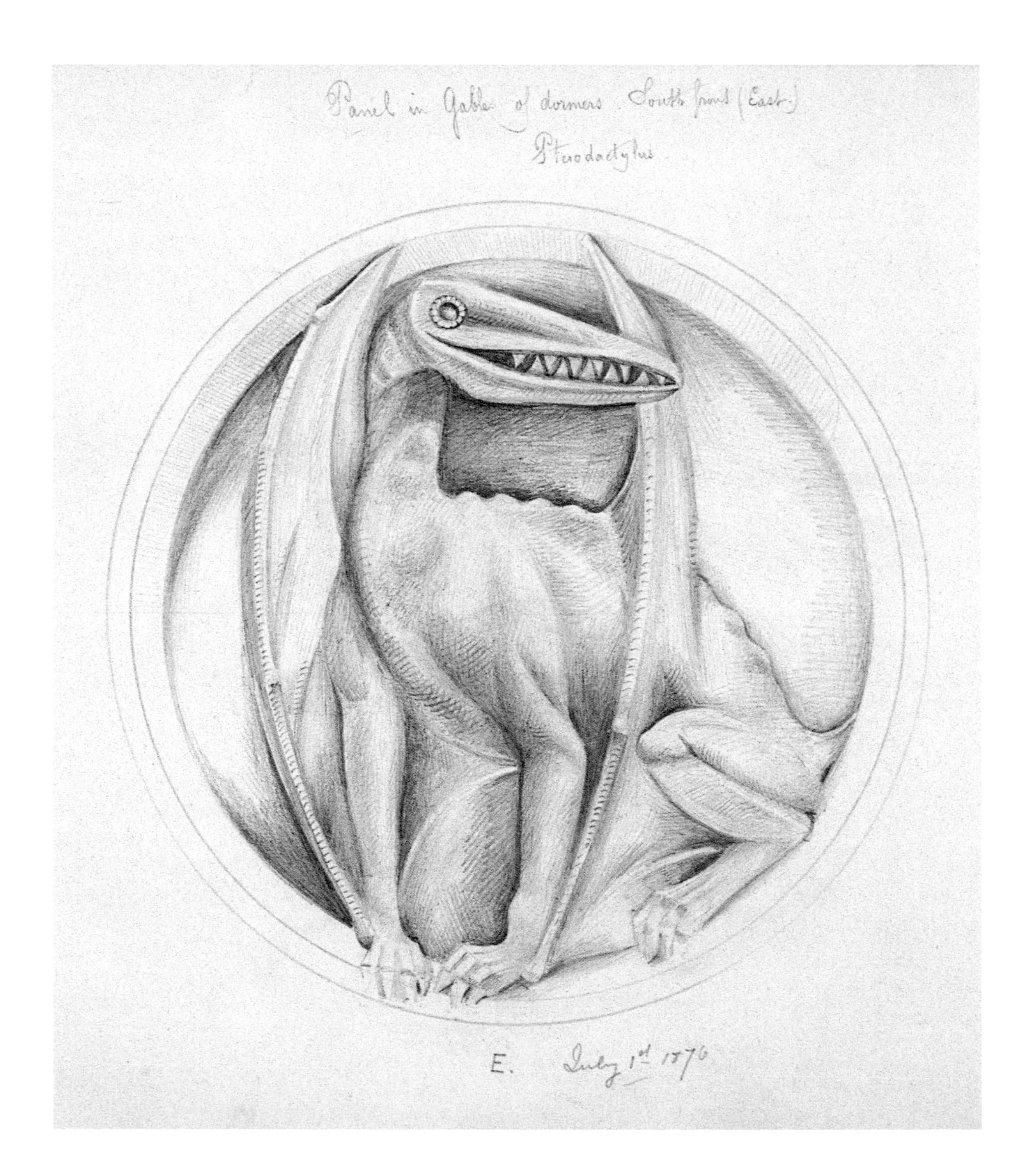
Panel in Gable of dormers. South front (East.)
Pterodactylus
E. July 1st 1876

Foreword

by Dr Michael Dixon, Director of the Natural History Museum, London

THE NATURAL HISTORY MUSEUM, LONDON IS ONE OF THE WORLD'S GREAT MUSEUMS. Internationally, its ever growing collections of over 70 million animals, plants, fossils, rocks and minerals are pre-eminent and illustrate the diversity of the natural world over time. Quite apart from their size and scale, their scientific and historical significance is immeasurable. The contain a very high proportion of 'type specimens', the actual specimens used in the first published description and naming of a species. As a result the collections are in frequent use by the Museum's own 300 or more scientists, and by around 8,000 visiting scientists who spend in total over 14,000 working days in the Museum on an annual basis.

The Museum's collections also contain an unrivalled natural history library and fine art collection with over 500,000 works of art, including superb watercolours of birds, flowering plants, mammals and insects, all chosen for their scientific accuracy as well as their artistic excellence. Consequently, they are widely used by historians of natural history and art. Their use is ever increasing as the Museum seeks to engage more with those who have an interest in the social, cultural and historical perspectives surrounding the Museum and its collections.

Voyages of Discovery throws a spotlight on the art and illustrative material generated from some of the most interesting and significant voyages of natural science discovery of the past 300 years. Each voyage led to highly significant new specimen collections and the generation of important new scientific knowledge. Many produced superb works of art, now housed in the Museum. The book describes the voyages, the well known and the less so, the dramatic and heroic, reproducing some of the rarest and most beautiful images created. Inevitably, there have been difficult choices to make about what to include and what to omit from the vast resource at the Museum. Much has never been seen by the public before, but these works deserve a large audience who I trust will derive great pleasure and fascination from these stories and images, just as generations of scientists and historians have done before.

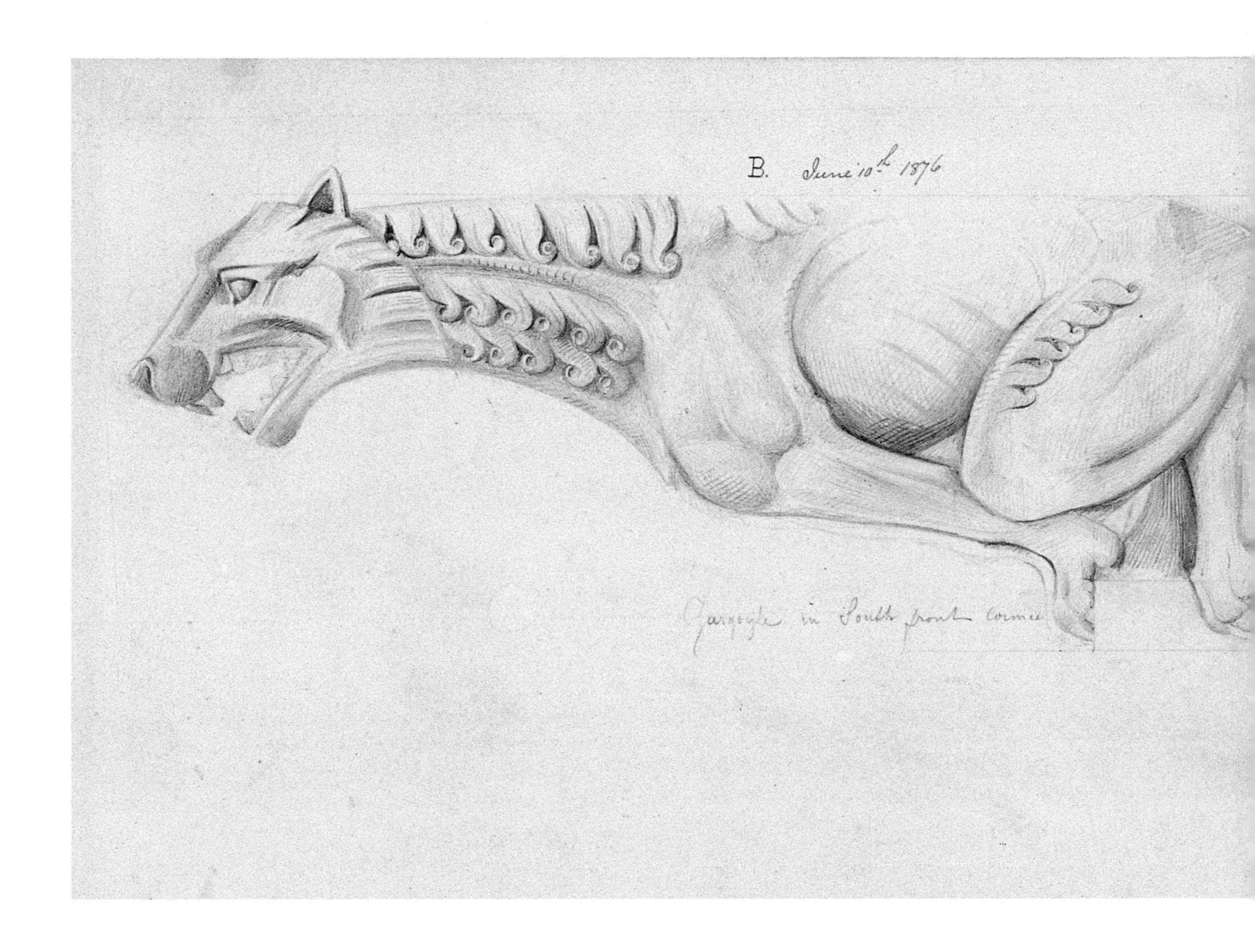

The building that houses the Museum is in itself a work of art, richly embellished inside and out with figures drawn from nature. These were designed by architect Alfred Waterhouse in the 1870s to reflect the content of the Museum's natural history collections, including both extinct and living species. Waterhouse's pencil sketches for the decorative terracotta *(above & pages 4, 9, 11)*, and his drawing of the finished building, dated 1876 *(page 12)*, form part of the institution's vast art collection.

Introduction

by Dr David Bellamy

I MADE MY FIRST VOYAGE OF DISCOVERY INTO THE WORLD OF NATURAL HISTORY OVER 60 YEARS AGO, scaling great heights to meet my first *Diplodocus* face to face. The heights? Well they were the steps up from the Underground to the front door of what is still for me one of the most exciting places in the world. Yes it has its rivals: diving uncharted coral reefs, pushing through pristine cloud forests or looking down a microscope at a drop of water from the ice at the North Pole. Each are voyages of personal discovery which hold that special excitement of erasing *terra incognita*. However, the voyage through the terracotta portals of The Natural History Museum is in many ways more awesome, for once within you are in the presence of the real giants of the past: not just the dinosaurs and pachyderms that grace the galleries, but the leviathans of natural history who made their voyages of discovery before the days of air travel, air conditioning and anti-malarials. It was their fortitude that laid the foundations not only of this museum but of taxonomy, genetics, evolution, continental drift, the theories that changed the way we think about ourselves, the planet on which we find expression and the role we play within the continuity of life. Their statues and portraits are there for ready reference, set amongst the glass cases or at rest in the stack rooms that hold the objects of their endeavour – household names like Cook, Banks, Linnaeus, Darwin and many more. Delve deeper and you will find the works of unsung heroes and heroines, without whose skill and devotion much of the vital information would have been lost. They are the artists who took the same risks as the superstars.

This fascinating book makes some amends for the capricious nature of historical record by allowing us to discover those missing links of human endeavour, itself part of the process of evolution. It is a time machine that focuses on just 300 years of the tumultuous story of life that is housed in this great building, a keystone period if ever there was one for it saw the turning of curiosity into science and curiosities into specimens, each of which gave new meaning to all our pasts and still forces us to ask questions concerning all our futures.

In 1699, just a few decades after it had been conclusively demonstrated that insects developed from eggs, not from mud, Maria Sibylla Merian travelled to Surinam to produce a series of paintings on the metamorphosis of butterflies,

including the food plants of larvae and imagoes. So good were her pictures that when Carl Linnaeus was producing his work describing all the animals then known to science, he included species she had recorded. Two Dutch artists, Hermann and de Bevere, shared the same honour, for it was their artistic expertise that allowed Linnaeus to write the flora of what we now call Sri Lanka. Thanks to the artistic talents of William Bartram, the plant and animal treasures of parts of North America were painted and documented, while in the Pacific, several artists played crucial roles on major expeditions: Sydney Parkinson on James Cook's *Endeavour* voyage with Joseph Banks and his naturalist Solander; George Forster on Cook's *Resolution* voyage; and Ferdinand Bauer, reputed to be the best of all natural history artists, on the *Investigator* with Matthew Flinders. The saga of Darwin on the *Beagle* is perhaps the most famous voyage of all. However without all the painstaking painting, drawing and cataloguing that had gone before, Darwin and his contemporaries Alfred Russel Wallace and Henry Walter Bates could never have set their findings within the context of natural selection.

The existence of this book and the treasure house of evidence it draws on is thanks to the vision of the British government of the late 18th century, a national lottery and the collections which Hans Sloane garnered in his lifetime. One of his most significant collections was made on his first voyage, to Jamaica, from whence he gave the world milk chocolate and a detailed account of the island's natural history, illustrated by a local artist, the Reverend Garret Moore. The end of the era – 300 years when science and art worked in symbiosis – came with the ultimate *terra incognita,* the challenge of earth's own inner space and the invention of photography. The scientists who were part of the voyage of the *Challenger* relied on a team of artists and photographers to make them famous as they plumbed the secrets of the oceans.

It is all too easy to dream about having been one of those early pioneers discovering things for the first time and to forget that in all those places, except the depths of the sea, people had been there long before discovering things for themselves. It is also easy to forget that every new voyage made by an individual is into *terra incognita.* So to be able to travel with the giants of the past is awesome indeed.

The Art of Nature

by Tom Lamb, Director of Books and Natural History Art Specialist, Christie's, London

The origins of natural history art can be traced back to the cave paintings of prehistoric man, developing through different cultures in a variety of forms – in the naturalistic imagery of Roman mosaics, in the Chinese silk flower paintings of the Han Dynasty, the decoration of Medieval illuminated manuscripts and in the development of flower painting in Europe, through to woodcut illustrations in natural history books of the 16th century. With few exceptions, early natural history imagery was formalised, even geometrical, in construction. However, by the late 17th century, the ideas of Newtonian science and philosophical advances encouraged a more natural approach to art with a new freedom of expression. This book takes up the challenge of this new expression, at the end of the 17th century, providing a glorious insight into the approaches and styles of some of the most important natural history artists of the last three centuries. In concentrating on the drawings from exotic voyages to the Americas and the Southern Ocean it is particularly exciting to see here some of the earliest European images of animals and flowers, especially considering the circumstances of the artists and the hardships they endured in travelling halfway around the world. The development of natural history drawing over this period of exploration reflects the change from the artist as an informed amateur to the artist as a trained professional, knowledgeable in plant and animal anatomy. That change is traced in these pages, from the work commissioned by Sir Hans Sloane and that of William Bartram, to the measured and careful scientific drawings of the *Challenger* reports. Between these two markers, *Voyages of Discovery* gives a fascinating insight into the diversity and development of the art of natural history drawing during this important period of exploration and discovery.

In the early 18th century, many voyages were privately financed and organised. One such expedition to Surinam by Maria Sibylla Merian brought about some of the finest artistic natural history images. The glorious collaboration of plant and insect life, sometimes inaccurate and archaic in style, but always compelling in their energy and composition, created an exoticism seldom surpassed. By the latter half of the 18th century, the organisation of expeditions improved considerably under government control, so that Cook's voyages to the Pacific, under the auspices of the Admiralty, employed established botanical artists such as Parkinson and Forster, who brought a more exacting and rigorous scientific

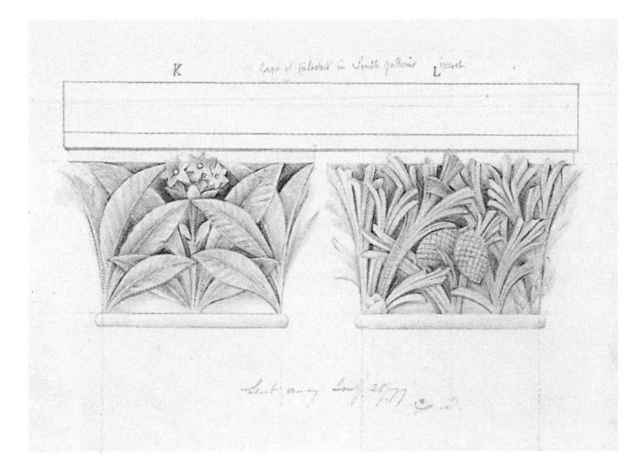

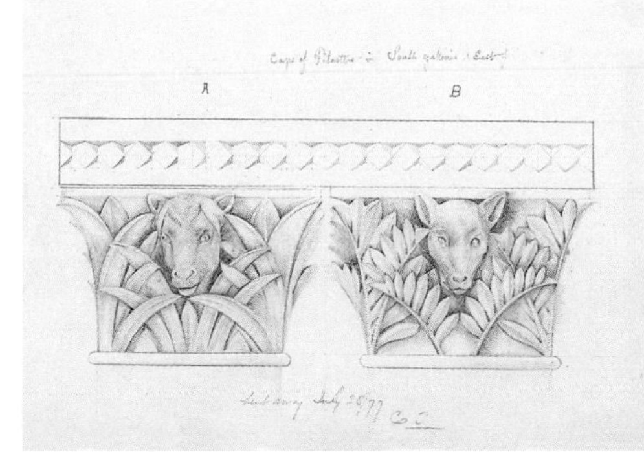

approach to natural history drawing by depicting, for example, both plant and seed as life-size, and explaining the environments in which they were found. The enormous achievements of Cook's voyages in the discovery and the opening up of the Pacific were matched by the naturalists who accompanied him in identifying, describing and drawing the new species of plants and animals that they found. This tradition reached a pinnacle with the work of Ferdinand Bauer who accompanied Flinders on the voyage of the *Investigator;* here a botanical artist of exceptional ability made detailed and impressive pictorial records of the plant and animal life around Sydney and along Australia's eastern coast. The standards set by naturalist-explorers such as Bauer, Parkinson and others in the latter years of the 18th century replaced those of the artistic amateur of the 17th and 18th centuries. In the early 19th century the role of the naturalist-explorer came to greater prominence. The work of Humboldt, Darwin, Wallace and Bates required the art of drawing to move ever closer to scientific accuracy. Such travellers were expected and required to be competent and careful artists, to record accurately and in detail the structure of plants or animals. The standards of scientific accuracy were now clearly set and were only to be matched, often surpassed, by photography in the 20th century.

Today the art of nature drawing is still alive, but how unfortunate that the present exploration of our planet and universe through aerial or space travel is not conducive to the pencil and sketchpad; technology and digital reality have largely taken over from artistic interpretation. Fortunate indeed are the collectors who can afford, and have the patience to wait for, the occasional great natural history drawing that only very rarely surfaces on the commercial art market. More fortunate are we to have the fine collections of The Natural History Museum to remind us of this glorious period of natural history exploration, thanks to Sir Joseph Banks, Sir Hans Sloane, Peter Collinson, the Admiralty and the many other patrons and institutions who saw fit to organise and collect these pictorial records, and that the British Museum was able to gather them for future generations to appreciate.

A Waterhouse. Mar 1876.

Voyage to Jamaica 1687–1689

Sir Hans Sloane

The 'Anchovy-Pear Tree' (opposite), now *Grias cauliflora*, of which Sloane wrote: 'The Fruit is by the Spaniards pickled and eaten in Lieu of Mangos, and sent from the Spanish West Indies to old Spain, as the greatest Rarity.'

LOVERS OF MILK CHOCOLATE WOULD PROBABLY NOT IMMEDIATELY SEE A CONNECTION between the object of their passion and the establishment of the British Museum. The curious link is a young Irish-born Protestant physician setting out on a long and distinguished medical career in late 17th-century London. In 1687 Hans Sloane was 27 years old, already had a well-established practice and was firmly ensconced in the medical and scientific society of the capital. Sloane's world was a turbulent one, politically, religiously and especially philosophically. There was still a widespread belief amongst savants that the 'correct' approach to the natural world was a totally detached and hypothetical one, resulting in interpretations of natural phenomena, including plants and animals, that frequently owed more to imagination than to fact. Consequently, most published accounts of natural history were still full of fictitious nonsense often based on fanciful travellers' tales brought back by uncritical observers from exotic parts of the world. But what was to become the world's most respected scientific society, the Royal Society 'for promoting natural knowledge', had been founded in 1660, the year of Sloane's birth.

The whole ethos behind the Society was the study of nature by careful observation and deduction, a rationalist approach fostered by such revolutionary thinkers as John Ray (1627–1705), the father of British natural history, and the physician and philosopher John Locke (1632–1704), both friends of Sloane. These 'new' men of science had no difficulty in accepting the standard religious view that the world and its populating plants and animals were the immutable creation of God, but saw the detailed observation, recording and interpretation of natural phenomena as a legitimate, indeed worthy pursuit. Sloane was firmly in this new mould and had been elected a Fellow of the Royal Society in 1685.

He was interested in all aspects of what we would now call science, from physics and chemistry through geology and palaeontology to natural history. But his first, and most abiding, love was for botany. This is not too surprising, for 17th-century medicine was intimately associated with the study of 'simples', the drug plants from which most medicines were obtained. Indeed, Sloane's botanical interest was fostered during his early years in London by the Chelsea Physic Garden, established by the Society of Apothecaries in 1673 specifically for this purpose – and later saved from

l. 7. 57.
Jun: 17. 1701

***An original Cadbury chocolate wrapper** (opposite)* from the 19th century, crediting Sir Hans Sloane with the recipe for milk chocolate. Sloane brought to England the beans of the cacao tree from Jamaica, where he observed that the Indians made a chocolate drink from the beans, which they took 'neat'. The Spaniards, however – who could drink chocolate as much as five or six times a day – liked to mix theirs with savoury spices.

closure by Sloane's financial intervention. But there were surely many new ones to be discovered and the New World had already provided the Old World with such useful products as the potato, maize, rubber, quinine and tobacco. So when the Duke of Albemarle was appointed Governor of Jamaica and offered the young doctor the post of physician to his family, Sloane jumped at it. The Duke's yacht, two merchant vessels and an escorting naval ship sailed from Portsmouth on 19 September 1687, stopping at Barbados for 10 days before arriving at Port Royal, Jamaica, on 19 December after what for the day was a relatively uneventful passage.

During the voyage, Sloane kept a meticulous journal in which he recorded all manner of observations on the daily shipboard routine, on natural phenomena and on the birds, fishes and invertebrates encountered along the way. He maintained the journal during his 15 months ashore, making notes on all sorts of topics including the weather, earthquakes, the island's topography and the behaviour of the local inhabitants, mainly escaped African slaves. But he also travelled extensively around the island, amassing and documenting a large collection of human artefacts, animals and particularly plants which, whenever possible, he pressed and dried for return to England. In dealing with the plants he was assisted by the first volume of John Ray's *Historia Plantarum*. In this book, Ray had tried to describe all the plant species then known and to arrange them first into major groups and then into smaller ones that he called 'genera', each with a synopsis of its characters as an aid to identification and naming. Although this was a great advance over previous works, it was still unwieldy and difficult to use and it would be another half-century before the Swedish scientist, Carl (Carolus) Linnaeus, provided botanists with a much simpler and user-friendly system.

Many of Sloane's samples, especially fruits, could not be preserved adequately, so he employed a local artist, the Reverend Garret Moore, to travel around with him and illustrate them while still in a fresh state along with many of the fishes, birds and insects that they encountered. The resulting illustrations, and most of the specimens including some 700 species of plants, eventually accompanied Sloane back to England. The specimens not yet drawn by Moore were illustrated by the talented artist Everhardus Kickius. Amongst the specimens he drew was chocolate, which Sloane had

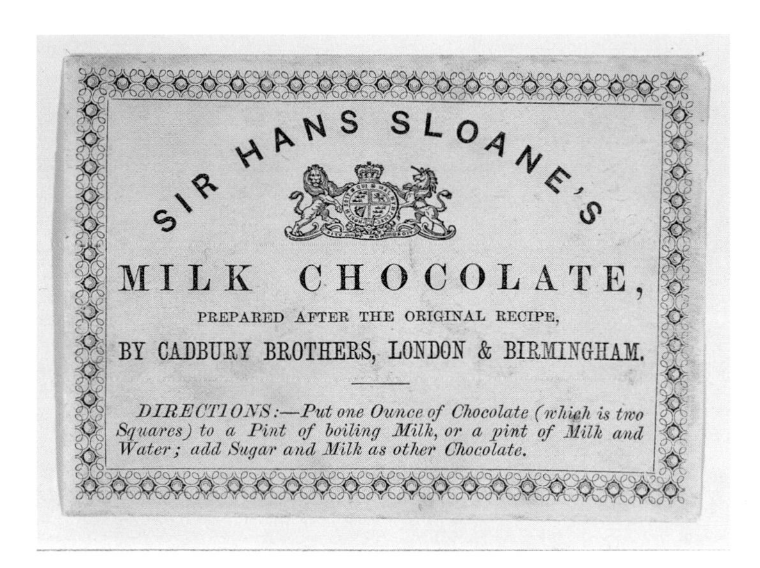

found was widely taken in Jamaica for its medicinal properties but was 'nauseous, and hard of digestion, which', he supposed 'came from the [chocolate's] great oiliness'. Sloane discovered that it was much more palatable when mixed with milk and his patented recipe brought him considerable income during his lifetime. In the 19th century, long after Sloane's death, the recipe was taken over by Cadbury, a name now more or less synonymous with milk chocolate in many parts of the world.

Sloane's main official duties in Jamaica were, of course, to look after the health of the Duke and his retinue, though he apparently also treated many others, including the retired buccaneer, Henry Morgan, by then respectable, knighted and the island's ex-Governor. But despite Sloane's ministrations, the Duke died in October 1688, still a comparatively young man. So the stay in Jamaica was cut short when the Duchess decided to return home, and Sloane's last duty for his erstwhile employer was to embalm him for the journey. But apart from the rather sad circumstances under which they were travelling, the homeward voyage throughout March, April and May of 1689 was rather more eventful than the outward one. For one thing, Sloane had a number of rather alarming live animals with him on board, including an iguana, a crocodile and a seven-foot-long snake. None survived the journey, however. The iguana inadvertently jumped overboard and was drowned; the crocodile died of natural causes; while the snake, which Sloane had had 'tam'd by an Indian, whom it would follow as a Dog would his Master ...', escaped one day from the large jar in which it was kept and was shot by one of the Duchess's alarmed domestic servants. Furthermore, the returning travellers were uncertain of the political situation they would find in England. Having originally left under the Catholic James II, but with dissension in the air, it was not until they approached the British coast that they learned from a fisherman that the Protestant William of Orange was now firmly on the throne.

Once back in London, Sloane picked up the threads of his interrupted career and reacquainted himself with a rapidly expanding circle of scientific acquaintances and correspondents. After almost four further years in the service of the Duchess of Albemarle he returned to private medicine and set up what was to become an extremely lucrative practice

A

VOYAGE

To the Iſlands

Madera, Barbados, Nieves, S. *Chriſtophers*

AND

JAMAICA,

WITH THE

Natural Hiſtory

OF THE

Herbs and *Trees, Four-footed Beaſts, Fiſhes, Birds, Inſects, Reptiles,* &c.

Of the laſt of thoſe ISLANDS;

To which is prefix'd An

INTRODUCTION,

Wherein is an Account of the

Inhabitants, Air, Waters, Diſeaſes, Trade, &c

of that Place, with ſome Relations concerning the Neighbouring Continent, and Iſlands of *America.*

ILLUSTRATED WITH

The Figures of the Things deſcrib'd,

which have not been heretofore engraved;

In large Copper-Plates as big as the Life.

By *HANS SLOANE,* M.D.

Fellow of the *College* of *Phyſicians* and Secretary of the *Royal-Society.*

In Two Volumes. Vol. I.

Many ſhall run to and fro, and Knowledge ſhall be increaſed. Dan. xii. 4.

LONDON:

Printed by *B. M.* for the Author, 1707.

The frontispiece to Volume I (opposite) of Sloane's two-volume account of the natural history of Jamaica and its neighbouring islands. In defence of the enterprise on which he had set out, he wrote in the preface: 'It may be ask'd me to what Purposes serve such Accounts, I answer, that the Knowledge of Natural-History, being Observation of Matters of Fact, is more certain than most Others, and in my slender Opinion, less subject to Mistakes than Reasonings, Hypotheses, and Deductions are ...'

in fashionable Bloomsbury, with clients including some of the richest and most prestigious figures of the day. As a result of this, and his marriage in 1695 to the heiress of a London Alderman and widow of a Jamaican estate owner, by the time Sloane died in 1753, at the age of 93, he was a very rich man. He was also very famous, not only as a physician and philanthropist, but also as a man of science and, particularly, as a collector of 'curiosities'.

Sloane had already started to collect botanical specimens in his youth and during his medical studies in London and in France. But as a result of his sojourn in Jamaica, his collections were enlarged enormously, and he continued to collect assiduously, especially West Indian material to add to his own in preparation for the account he intended to publish detailing his finds in Jamaica. In 1696, he published his *Catalogus Plantarum*, a relatively simple 232-page list documenting all the plants he had found on the island. In doing so he set a high standard for such works, referring carefully to all available earlier sources of information in order to try and avoid the nomenclatural confusion that was all too common in the days before the now universally accepted binomial system was introduced by Linnaeus in the mid-18th century. But the production of the full account of his Jamaican experiences, his *Natural History of Jamaica,* took much longer.

The first volume, concerned mainly with plants, was published in 1707, while the second, which included Jamaican zoology, did not appear until 1725. Both were illustrated with engravings by Michael van der Gucht – one of the best in his field – based on original drawings by the Reverend Moore in Jamaica and, more usually, by Everhardus Kickius back in England. Although Sloane published little else of a scientific nature, these volumes so enhanced his reputation that the Royal Society, for which he had already served as Secretary from 1693 to 1713, elected him as President to succeed Sir Isaac Newton who died in 1727. Sloane was to retain this post until 1741 when he resigned it immediately before he gave up his medical practice at the age of 81 to retire to Chelsea.

But even in retirement and old age, Sloane continued a wide-ranging correspondence with scientists and received a stream of visitors, both from Britain and abroad, to view his famous collections. Indeed, it is said that during his long life, Sloane knew personally, or by correspondence, everyone who was worth knowing in science, and particularly in botany.

Illustrated pages *(opposite)* from Volume 2 of Sloane's *Natural History of Jamaica*, published in 1725. Although the majority of engravings in both volumes were of plants, Volume 2 included 41 zoological plates. Volume 1 also included a fold-out map of the island of Jamaica *(overleaf)*.

By the time of his death, the original basis for the collection – the herbarium of pressed plants – had grown by Sloane's own purchases and his acquisition of the entire collections of other botanists to fill no fewer than 265 huge leather-bound volumes, a nucleus of eight of them devoted to material resulting from the Jamaica period.

Famous though the herbarium was, and still is, it has not been without its critics. In 1736, the 29-year-old Carl Linnaeus, soon to become a far more celebrated botanist than Sloane, visited the 76-year-old physician. Linnaeus was understandably deferential to the older man face to face. But when he returned to Sweden, he was openly critical of what he considered to be a chaotic way of keeping specimens in permanent bindings. For Linnaeus was already employing the later universal practice of storing his specimens on single unbound sheets that could easily be re-ordered to accommodate changes in the classificatory system used and to allow new material to be incorporated.

But Linnaeus would certainly have been impressed by the sheer volume of Sloane's collections. Apart from the herbarium sheets, a further 12,500 'vegetables and vegetable substances' occupied thousands of small glazed boxes in 90 drawers in five cabinets. But Sloane cast his collecting net much, much wider than botany. His zoological collections, which essentially started with the material he brought back from Jamaica, included almost 6,000 shells, more than 9,000 other invertebrate specimens (half of them insects), 1,500 fishes, about 1,200 birds, eggs and nests, and more than 3,000 vertebrate specimens ranging from stuffed whole animals, through hundreds of skeletons – including that of a young elephant and a whale skull five-and-a-half-metres (18 feet) long – to a grisly assortment of bizarre human 'curiosities'.

Even these represented only a fraction of his entire hoard, for Sloane's tastes and collections embraced thousands of fossils, rocks, minerals, ores, metals and precious or semi-precious stones, all either in their natural unworked state or incorporated in jewellery, ornaments or practical objects. His antiquities and ethnographic collections ranged from the classical to the then recent, from the Old World, to the New World and the Orient, and included 32,000 medals and coins. His collection of 300 or so examples of more or less conventional art was neither particularly big nor remarkable, though it included works by some notable artists such as Albrecht Dürer (1471–1528). But the library was perhaps the

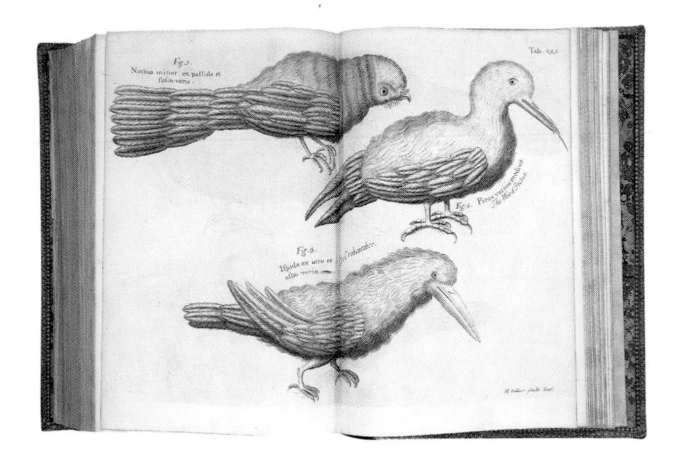

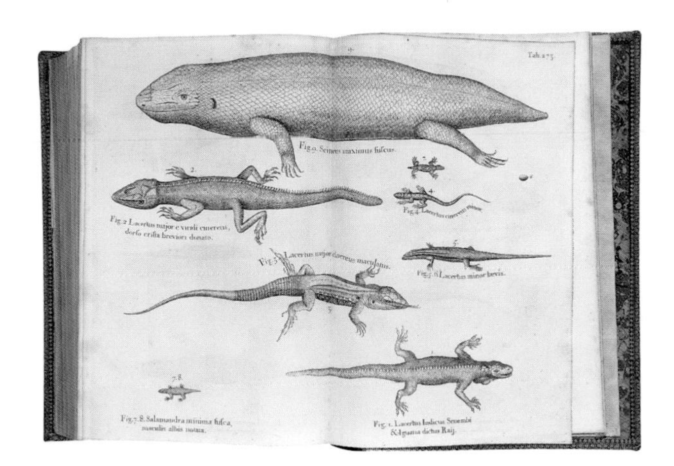

crowning glory. Containing almost 50,000 bound volumes of published works, many of them lavishly illustrated, as well as an enormous collection of manuscripts and drawings on all manner of subjects, it was undoubtedly one of the most comprehensive libraries of the time.

During his old age, Sloane understandably became increasingly concerned about the fate of the collections after his death. He was most anxious that they should be as accessible as possible to anyone with a genuine interest in them, for whatever purpose. After deciding that no existing institution – the Royal Society, College of Physicians or the Ashmolean Museum in Oxford – would be a suitable repository for the collections, Sloane determined to leave them to the nation. Apart from the stipulation that they should be properly housed, maintained and made accessible, his only requirement was that his two daughters should be paid a total of £20,000 for them, much less than the £100,000 he estimated he had invested in his collections.

Sloane died on 10 January 1753 and his will inevitably caused a good deal of controversy. But the net result was that on Thursday 7 June 1753, the British Museum was established by an Act of Parliament, with Sloane's collections, and two smaller ones purchased to join it, forming the nucleus of the new institution. The government chose to raise the necessary funds by a national lottery, not an unusual practice in the 18th century. From these relatively modest beginnings developed the present renowned institution with its seven million-plus human artefacts in Bloomsbury, over 67 million items in the British Library and 68 million natural history specimens at The Natural History Museum in South Kensington – including Sloane's specimens of the cacao plant, the ultimate source of milk chocolate.

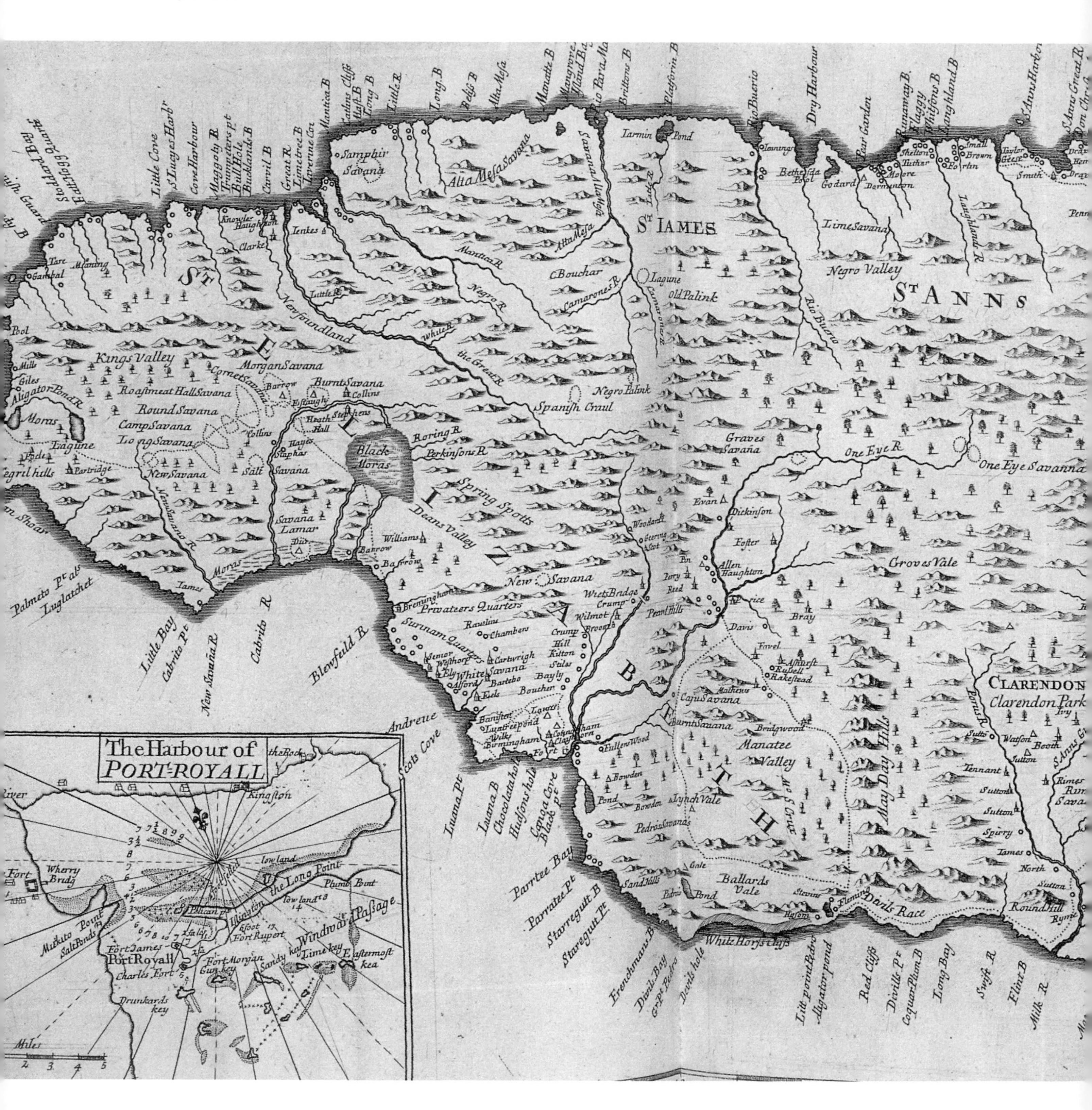
The Harbour of PORT-ROYALL
St IAMES
St ANNS
CLARENDON
Clarendon Park
Kings Valley
Morgan Savana
Cornet Savana
Burnt Savana
Roastmeat Hall Savana
Round Savana
Camp Savana
Long Savana
New Savana
Salt Savana
Savana Lamar
Black Morass
Spring Spotts
Deans Valley
New Savana
Privateers Quarters
Surinam Quarters
White Savana
Spanish Craul
Negro Palink
Old Palink
Lagune
Graves Savana
Groves Vale
One Eye R
One Eye Savanna
Negro Valley
Lime Savana
Alta Mesa Savana
Samphir Savana
Camarones R
Manatee Valley
May Day Hills
Cajus Savana
Burnt Savana
Lynch Vale
Pedros Savanas
Ballards Vale
Devils Race
White Horss cliffs
Round Hill
Dry Harbour
Rio Bueno
Pear Garden
Runaway B
Little Cove
Cove Harbour
Maggoty B
Great R
Mantica R
Negro R
Newfoundland
the Great R
Roring R
Perkinsons R
Palmeto Pt als Luglatchet
Little Bay
Cabrito Pt
New Savana R
Cabrito R
Blewfeild B
Scots Cove
Andrew
Luana Pt
Luana B
Chocolata hole
Hudsons hole
Canoa Cove
Black Pt
Parrtee Bay
Parratee Pt
Starregutt B
Starregutt Pt
Frenchmans B
Devils Bay
Devils hole
Alligator pond
Red Cliffs
Devills Pt
Long Bay
Swift R
Flint B
Milk R
Kingston
the Long Point
Plumb Point
Windward Passage
Fort Rupert
Port Royall
Charles Fort
Fort James
Fort Morgan
Gun key
Sandy key
Lime key
Eastermost key
Drunkards key
Muskito Point
Salt Ponds
Wherry Bridg
Fort
Miles

A NEW MAP OF THE ISLAND OF JAMAICA
The Characters Exprest in the Map are
For a Towne
For a Church
Sugar Work
Indico work
Cotton & Provisions
a Cacao walk
Crawles for Hoggs & Pens for Cattle
St MARIES
Rio Hoa Savana
Mount Maria
Ballards Valley
White Cliffs
Pedros Cockpitts
Old woman's Savana
Anchovy Valley
Rio de Flora
Magotty Savana
Spring Savana
St THOMAS IN THE VALE
BLEW MOUNTAINS
St GEORGES
St THOMAS
St DOROTHYS
St JOHNS
Spanish Town
St KATHRINES
St Andrews
Liguanee
Kingston
St DAVIDS
Port Royall Harbour
Port Royall
The Long Point
Passage Fort
Modifords Creek
Old Harbour
Carlile
Portland Cay
Sandy bank
Pidgeons I
Port St Thomas
Yallah Point
Yallah Bay
Rio Grande
Seamans Valley
Two Mile Rock
Plantain Garden
Miles
Leagues
5 10 15 20 25 30

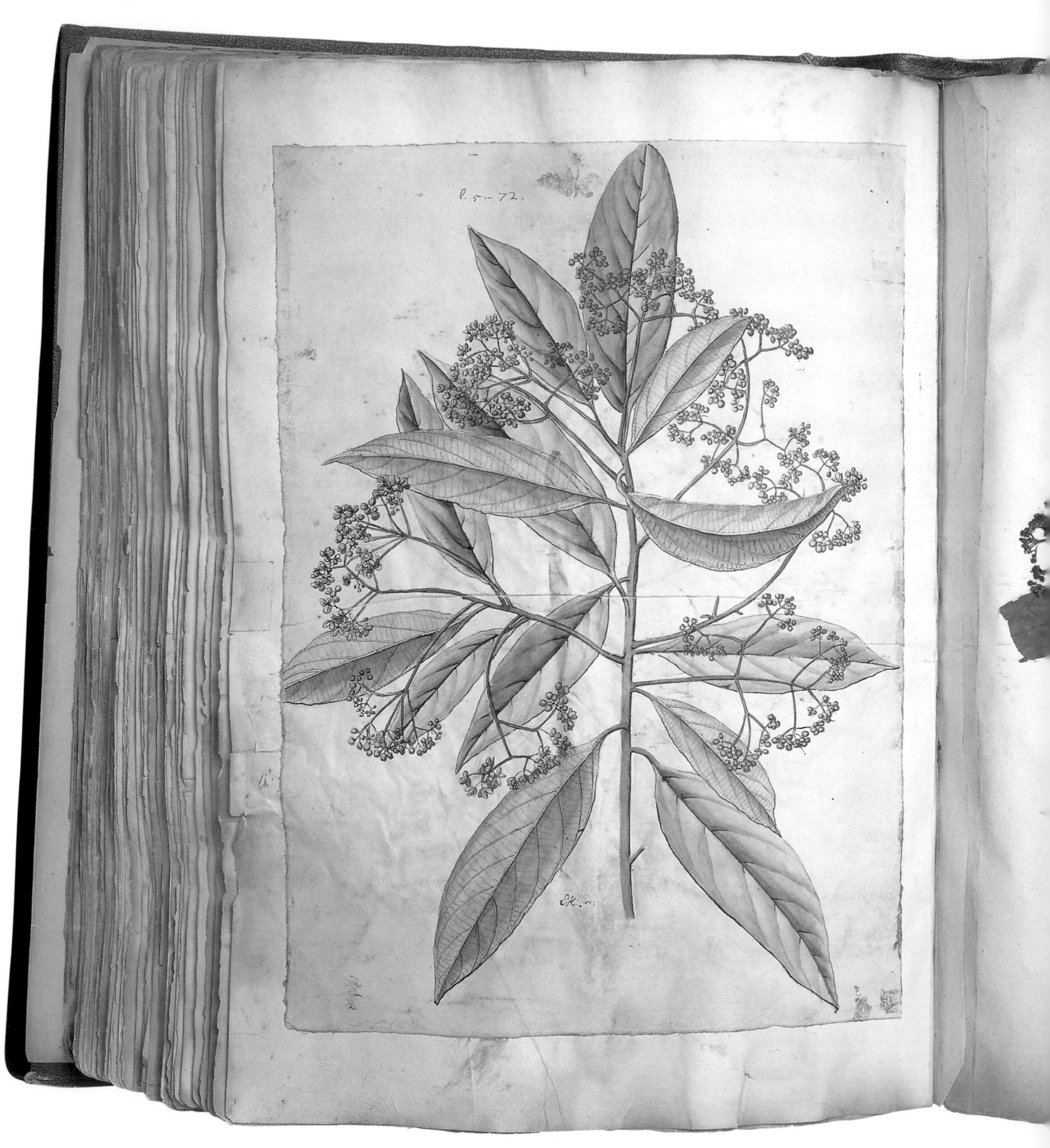

***A leather-bound volume** (left),* part of Sloane's famed Jamaica herbarium of pressed specimens and illustrations. He likened the shrub on these pages to the bay tree familiar back in Europe, and called it *Laurus folio longiore, flore hexapetalo racemoso, fructu humidiore.* Under the Linnaean system, it is now called *Nectandra antillana*, commonly shingle wood, sweet wood, or yellow sweet wood. Before Carl Linnaeus standardised plant nomenclature with his binomial, or two-part, system for names, Sloane and other botanists had used long descriptive Latin phrases. This inevitably led to enormous confusion, made worse because many botanists did not check earlier work to see whether a particular plant had already been described. Sloane, however, was meticulous in referring to his predecessors, using the best available classifications, particularly those of the British botanist John Ray.

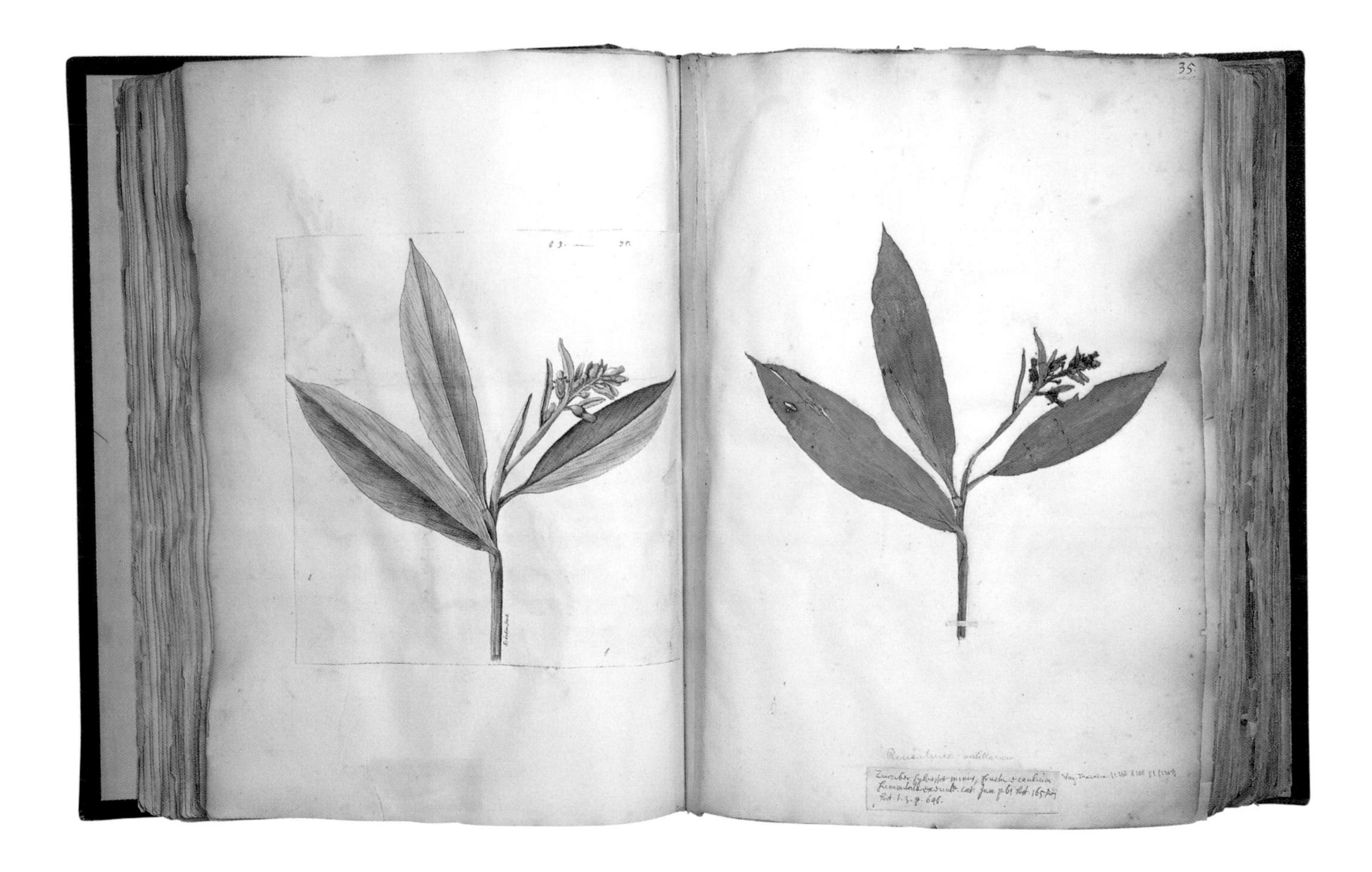

Zinziber sylvestre minus, *fructu è caulium summitate exeunte* or 'wild ginger' *(above),* as Sloane described it – now known as *Renealmia antillarum.* According to Sloane, Jamaicans held the plant's medicinal properties in high esteem: 'The Root bruised, and applied as a Poultice in Cancers ... is reckoned a very extraordinary and admirable Medicine, and if one will give Credit to the Relations of Indians or Negroes, is a never failing Remedy in those desperate Cases,' he reported. In his description of what he dubbed the 'Bastard Locust-Tree' *(opposite)* – now known as *Clethra occidentalis,* or soapwood – he notes that the berries are edible: 'The Pulp is sweet, white, mealy, including a hard, brownish, black Stone, bigger than a Pepper-Corn, and much like it ... The Berries are ripe in August, then fall off the Trees, under which they are gather'd and carried to Market, being eaten and thought a pleasant Disert.'

l. 6. — 99.
E: Kikus fec.
Mayz. 1701.

Sloane's labelling of this tree *(below & opposite)*, found on the north side of the island of Jamaica, was *Laurifolia Arbor flore tetrapetalo*. The straight branches were covered with a smooth, dark brown bark, enclosing a white wood. The leaves, which grew in random arrangement 'at the Ends of the Twigs', were 5 centimetres (2 inches) in length and 2.5 centimetres (1 inch) across at their widest part, and 'smooth, shining, thick ...' Clusters of pale yellow, four-petalled flowers (hence *tetrapetalo*) gave way to small, round fruits 'not so large as a Pepper-Corn, but very elegant and pretty', Sloane concluded. The tree's modern name is *Erithalis fruticosa*.

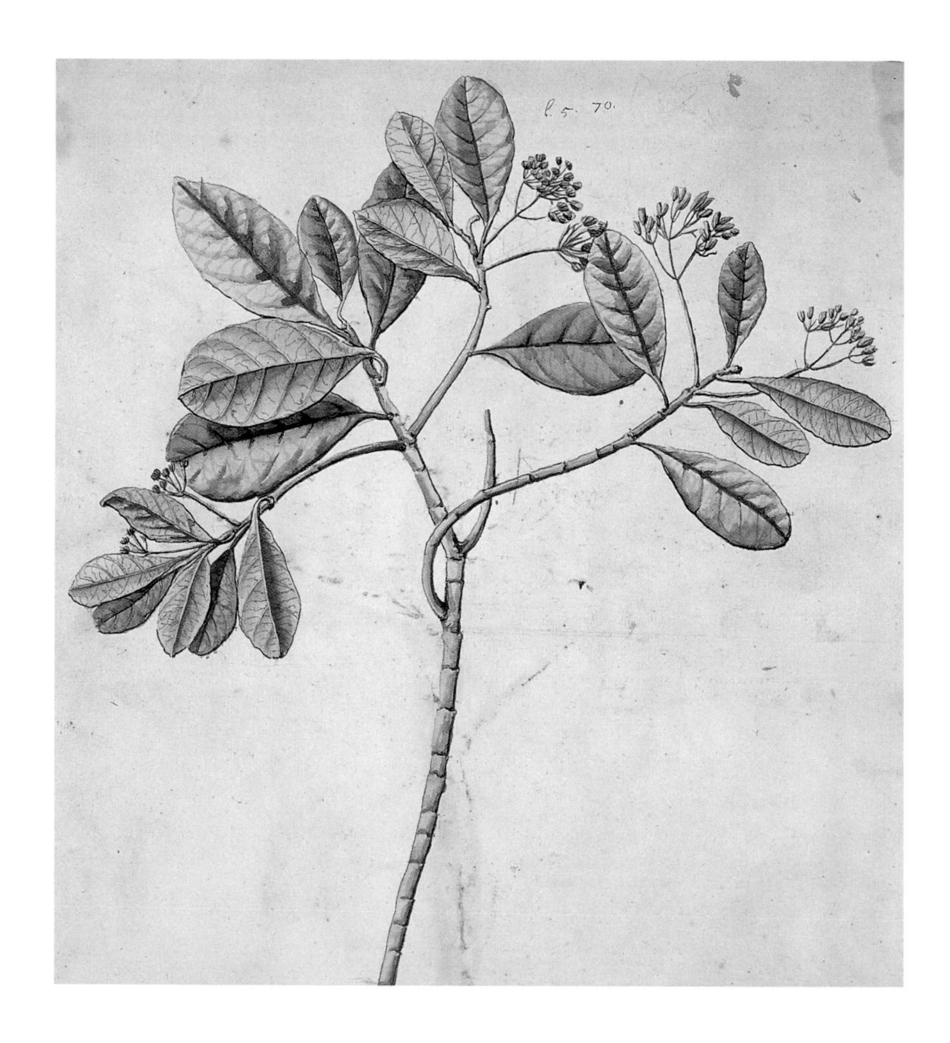

Laurifolia arbor flore tetrapetalo, fructu racemoso ro-
-tundo cannulato & coronato. Cat. Jam. p. 136. Hist. vol.
2. p. 20. Raij. Hist. t. 3. D, p. 86.
Erithalis fruticosa Linn.

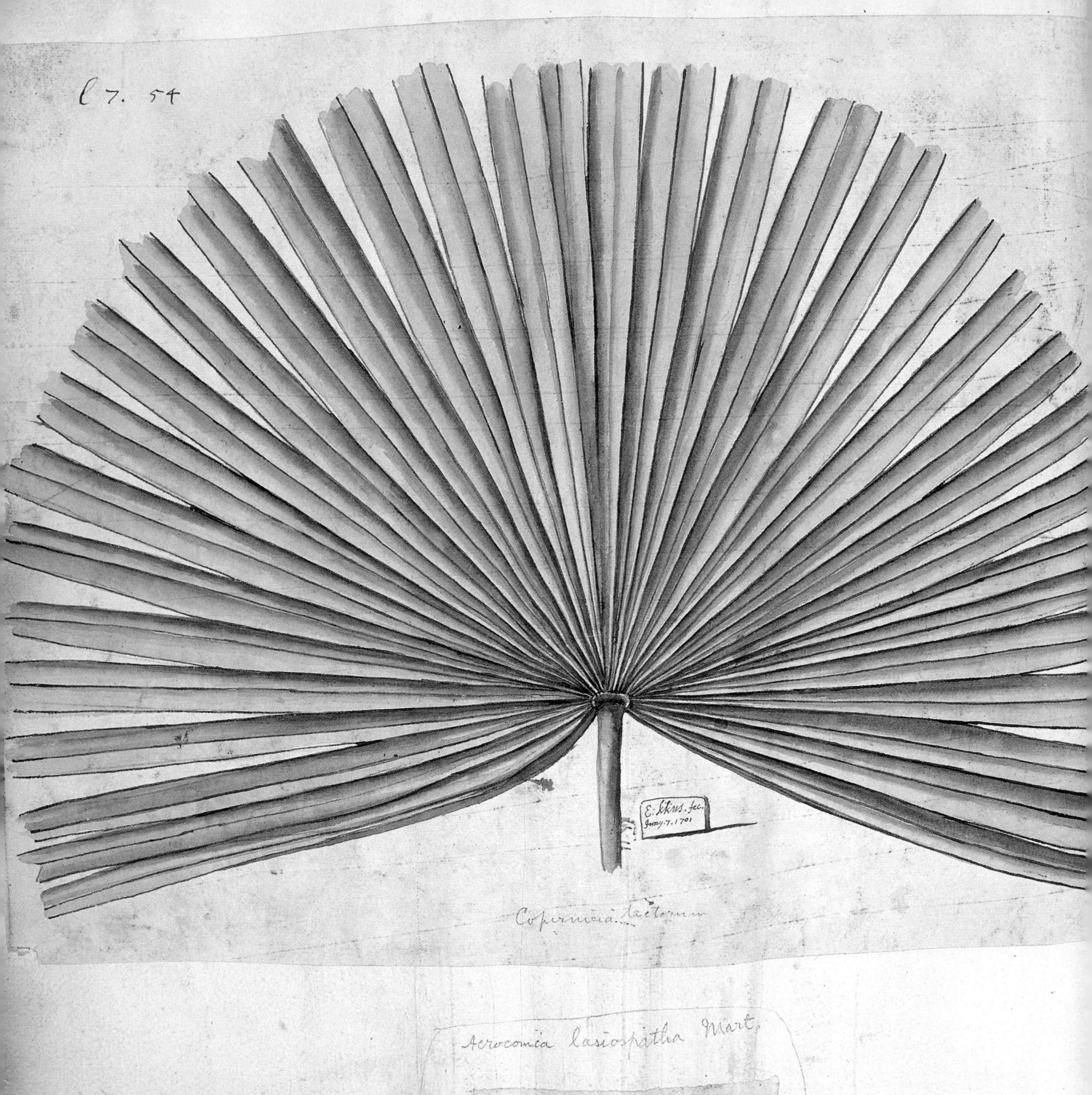

Palma tota spinosa major fructu
pruniformi. cat. Jam. p. 177. hist.
vol. 2. p. 119. Ray. hist. p. 1363?

Sloane named the tree from which these leaves came *(opposite & below) Palma Brasiliensis prunifera folio plicatili seu flabelli formi caudice squamato*, classifying it as a palm belonging to his group of 'Pruniferous Trees'. It is now *Acrocomia spinosa*. The tree provided raw material for numerous purposes in Jamaica, particularly thatching (hence its common name, 'thatch'). Its bark was used for making boxes and baskets, and its wood for bows, clubs, darts and arrowheads. The leaves provided fans for fanning the fire instead of bellows, and they were also used to cover salt to keep it dry. In times of scarcity, the roots and berries could be eaten.

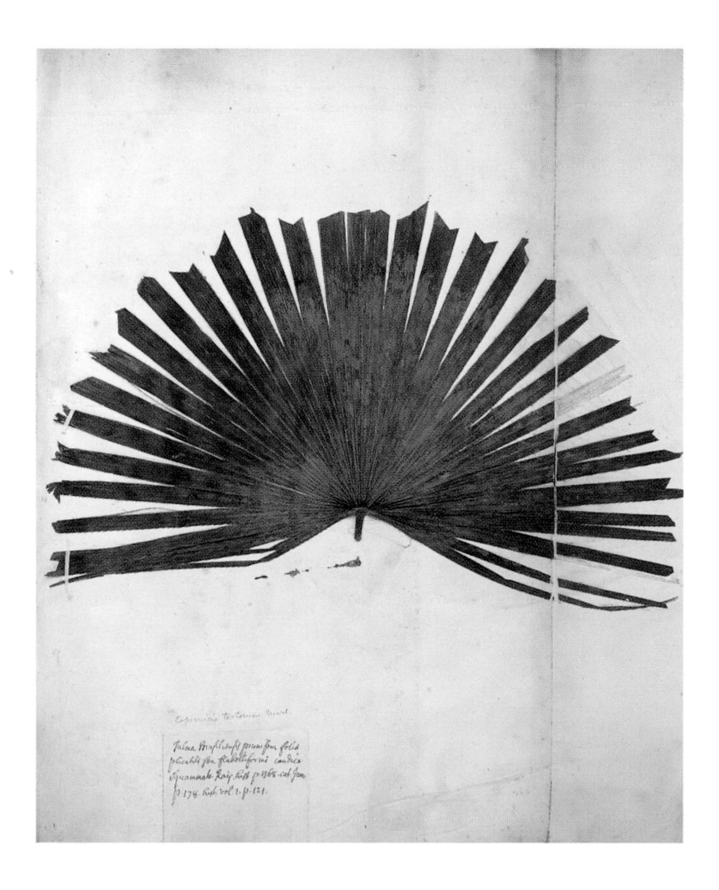

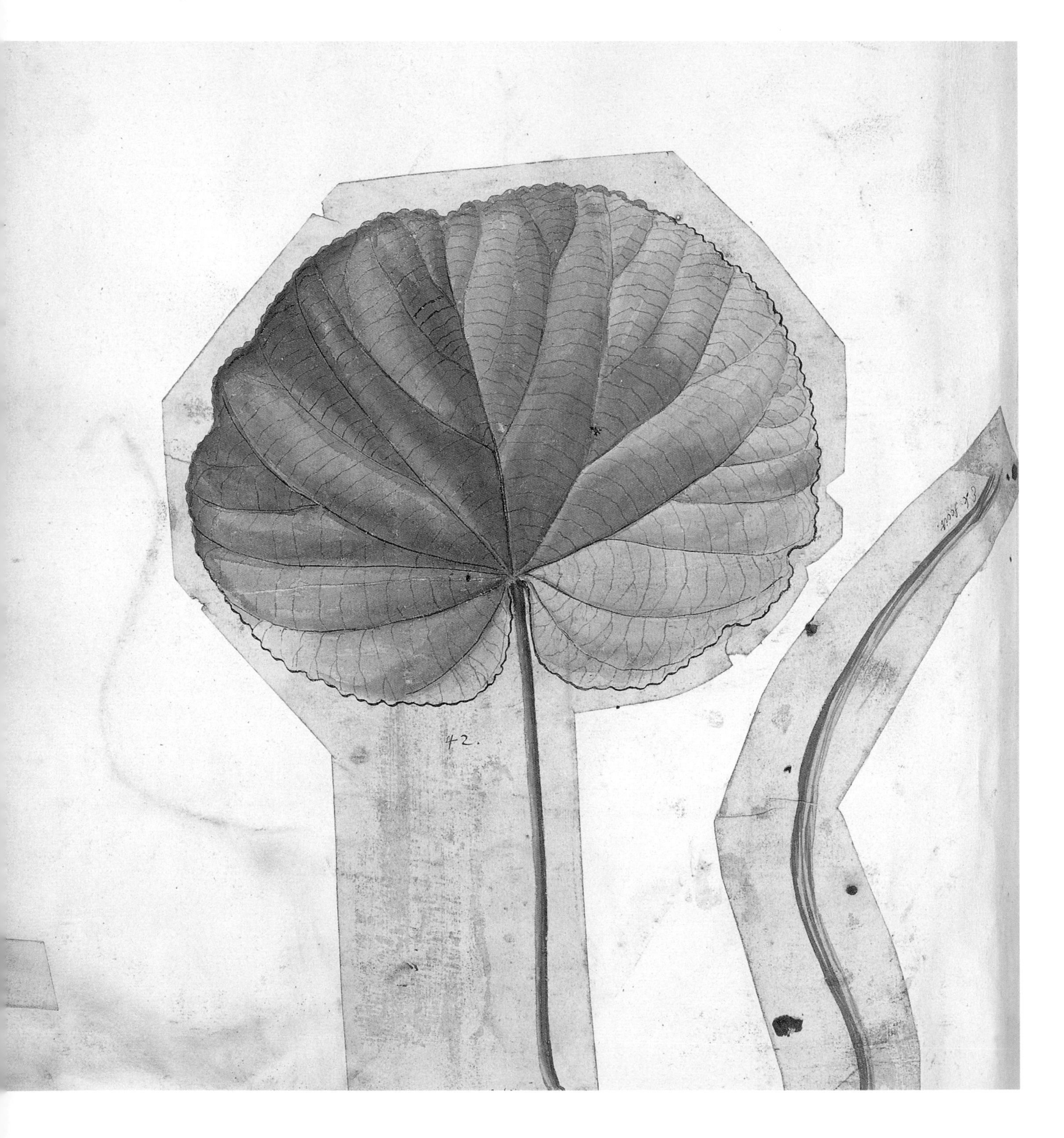

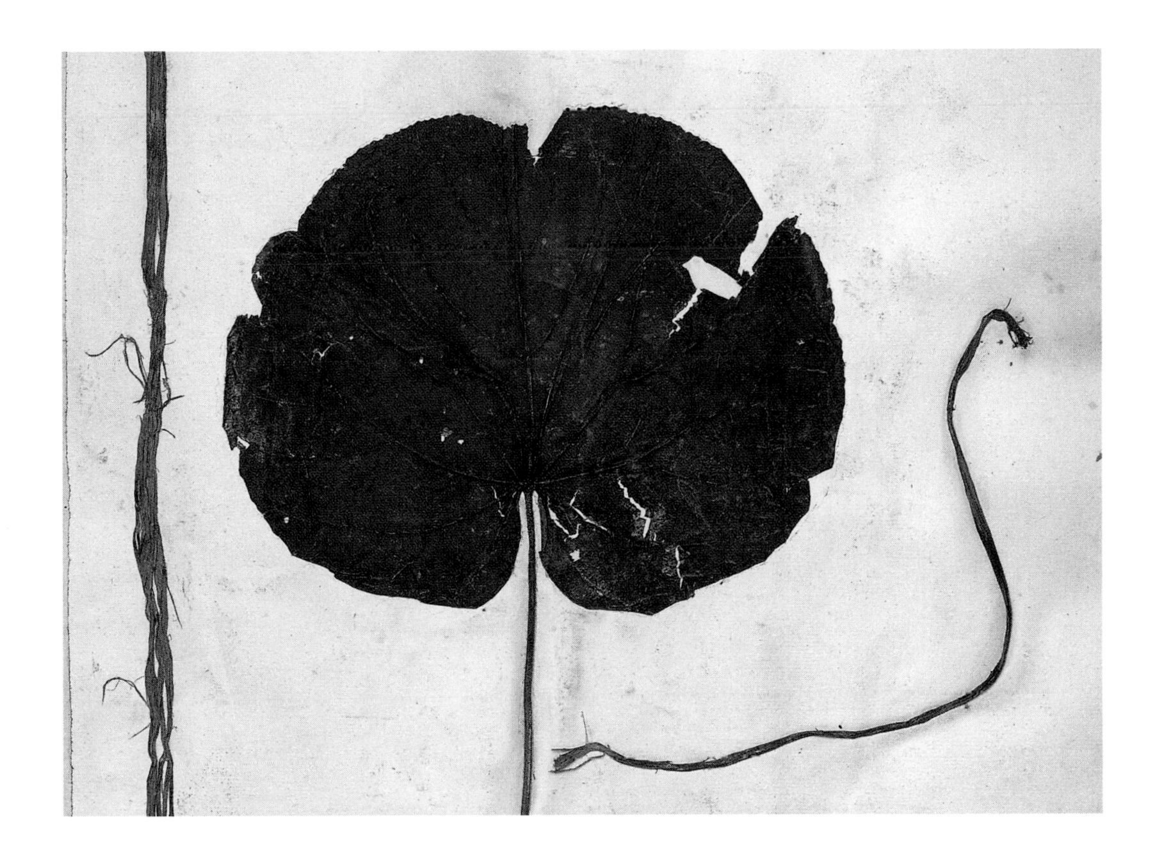

Sloane called this plant *(opposite & above) Malva arborea, folio rotundo, cortice in funes ductili, flore miniato maximo liliaceo* or the Mahot or Mangrove-Tree. Its modern name is *Hibiscus elatus*, commonly known as cube bark, blue mahoe or mountain mahoe. The tree had rounded leaves (hence *folio rotundo*) and flowers with five petals enclosing a red 'Pestle or Stylus of the same length, on which are many Stamina, the whole Flower looking like a red Lilly.' The bark apparently had a practical use. Quoting another source, Sloane says that 'of the more gross [bark] is made Cords [ropes] of the other Britches, for the Negroes and Slaves.'

TYPE SPECIMEN
Erythrodes plantaginea (L.) Fawc. & Rendl.
Satyrium plantagineum L.
J.D. Ackerman 1987
Erythrodes plantaginea

This orchid, *(opposite & left)* was described as *Orchis elatior latifolia asphodeli radice, spica strigosa,* and grew, Sloane wrote, in the woods on Mount Diablo. Its stalk was about 45 centimetres (18 inches) high, with alternate leaves on either side which were 7.5 centimetres (3 inches) long and 4 centimetres (1½ inches) wide at their widest point. A slender flower rose from the top of the stalk: 'the Petiolus [stalk] of the Flower is crooked, the Spur [nectar-bearing part] blunt, the Labellum [posterior petal] small, and the Galea [hood] large, and divided as others of this kind.' The plant is now called *Erythrodes plantaginea.*

Edwd: Kickius fecit.

Many pre-Linnaean names survived into modern botanical nomenclature. For example, Sloane begins his diagnosis of this sedge *(opposite & above)* – which he placed in his group of 'Herbs with grassie Leaves' – *Cyperus longus odoratus*. This was exactly the same wording used by John Ray to describe the closely related English galingale. Under the Linnaean system, this species became *Cyperus odoratus*, while the English galingale became *Cyperus longus*.

Apocynum erectum fruticosum *flore luteo maximo & speciosissimo*, or the 'Savanna Flower' *(below)*, a plant now known as *Urechites lutea*. Sloane saw it growing 'in the Savannas every where' and noted that it was 'in Flower most part of the year, making a very pleasant sight'. Perhaps more significant was the source of the milk chocolate drink Sloane devised – the cacao tree *(opposite)*, *Theobroma cacao*. Writing of the beans, he said: 'The Nuts themselves are made up of several Parts like an Ox's Kidney, some Lines being visible on it before broken, and is hollow within, its Pulp is oyly and bitterish to the Taste ...'

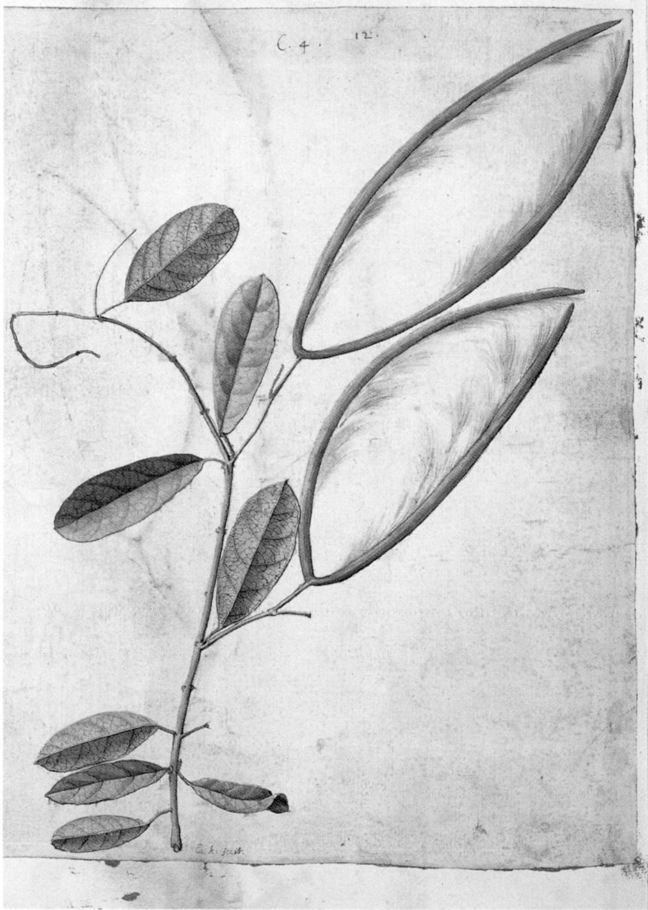

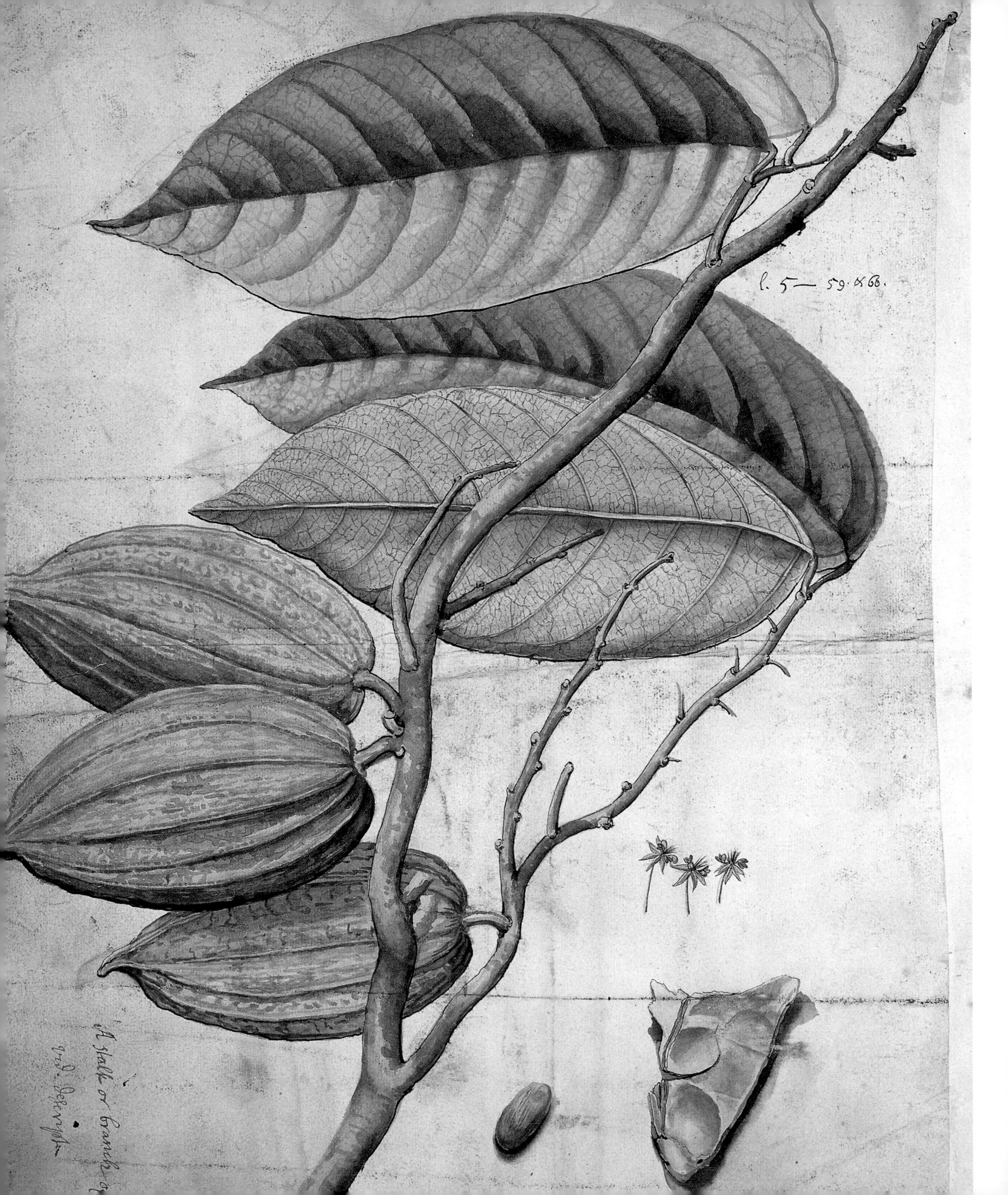
A stalk or branche

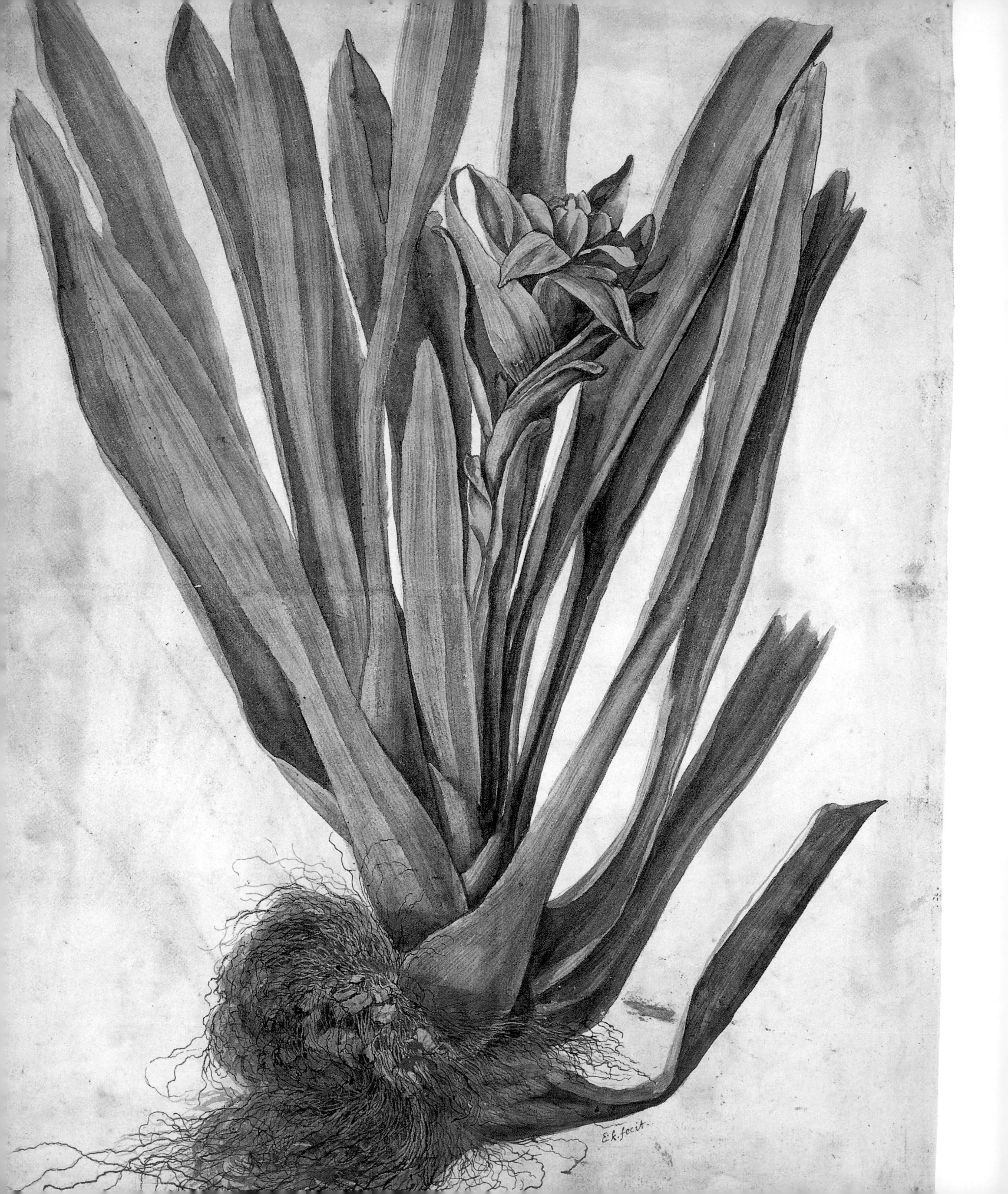
E.k. fecit.

Viscum Caryophylloides *maximum capitulis in summitate conglomeratis* was Sloane's diagnosis for this bromeliad *(opposite & left):* 'This has a great many long, dark, brown, small filaments, threads, or fibrils, which take fast hold of the Barks of Trees, to which it adheres, when all united making an oblong Root ...' At the top of the stalks, a rose-like cluster of green and reddish leaves, covered with a 'glewy mucilage', enclosed the seedheads. The plant grew on large, old trees but, Sloane wrote, it could also take root on the ground if it happened to fall there. Its modern botanical name is *Guzmania lingulata.*

***A specimen** (above) **and drawing** (opposite)* of a shrub which Sloane named *Caryophyllus spurius inodorus, folio subrotundo scabro flore racemoso hexapetaloide coccineo speciosissimo*, but is now *Cordia sebestena*, the red or scarlet cordia. It had an upright growth habit, Sloane wrote, rising by means of several trunks, covered in 'Clay colour'd Bark', to a height of 2.4 or 2.7 metres (8 or 9 feet). The leaves were almost round, harsh to touch, and dark green in colour, while the 'many and large' flowers were a delicate scarlet. 'The Fruit I never found in Perfection,' Sloane remarked, ending with: 'It grew on a rocky Bank over Mr. Batchelor's House near the black River Bridge, and made there a most pleasant Sight.'

l. 5.-71

Cyperus minimus Linn.
Gramen cyperoides minimum, spicis pluribus compactis ex oblongo-rotundis. Cat. Jam. p. 36. Hist. 120. Raij. Hist. t. 3. p. 625.
Voy. Jamaica 1:120 t.79 f.3 (1707)

This tiny sedge *(opposite & above)* was described by Sloane as *Gramen cyperoides minimum*. It grew to only about 7.5 centimetres (3 inches) tall. The specimen is so well preserved that even the fine roots can still be seen. Sloane was puzzled by the role played by grasses and sedges in the Jamaican ecosystem: 'I doubted very much whether I should find in the American Islands any Grasses, at least in Plains as our Fields in Europe, but I found many grassie Plains, and in them Kinds of Grasses analogous to those of Europe ... What the design of Nature was in their Production seems hard to discover, for in these Islands they had no large Fourfooted Beasts but one, till Europeans landed there, unless it be said that as Corn with greater Seeds are for Man's Nourishment, so these were appointed for the Food of Birds and Insects, which feed on them and their Ripe Seeds.'

A specimen,** (below)* ***and drawing *(opposite)* of *Cecropia peltata*, alternatively known as snake wood or the trumpet tree. Sloane entitled it *Yaruma de Oviedo*, and reported that it was widespread in the woodlands of Jamaica and the other islands of the West Indies. It had several practical purposes: its leaves and juicy pith were used by 'Indians and Negroes' to dress their wounds, and in Brazil, Sloane noted, its wood was used as firesticks. Another plant, shown here as a specimen *(right)* and a drawing by Kickius *(below right)* dated 31 May 1701, was compared by Sloane to the familiar holly of Europe because of its prickly leaves. Its modern botanical name is *Drypetes ilicifolia*. Its common name is rosewood.

From the third herbarium volume, a plant *(below)* described by Sloane as *Arum saxatile, repens, minus, geniculatum & trifoliatum*, but now called *Philodendron tripartitum*. A climber, it had stems with a spongy pith and milky sap. Along the stems were thick joints, and out of each grew 'five or six Clavicles [suckers] ... [with which] it takes hold, and sticks very close, and fast to any Tree it comes near.'

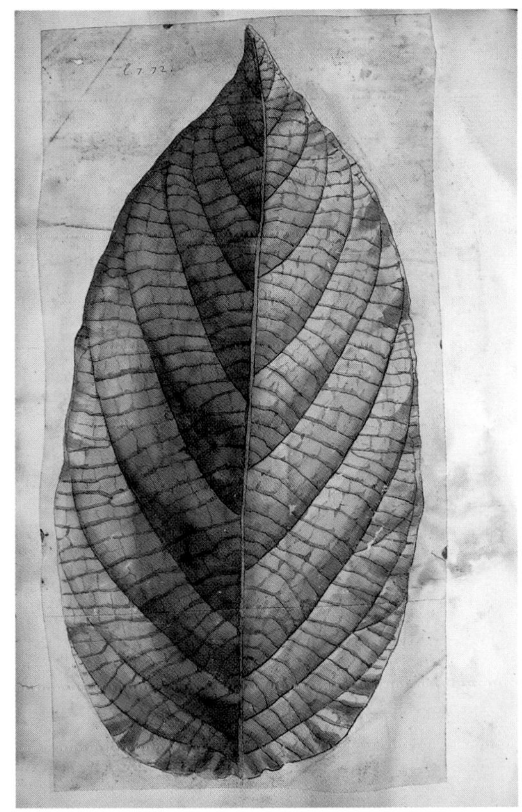

A leaf from a tree *(above)* which, according to Sloane, was *Prunus racemosa, foliis oblongis hirsutis maximis, fructu rubro,* or the 'Broad-leaved Cherry-Tree'. Faced with an unidentified plant, Sloane named it according to what he knew, placing it in the genus *Prunus* familiar back in Europe. The leaves were 'hoary, corrugated like Sage or Fox-glove, woolly [hence *hirsutis*], and of a fresh green Colour'. The tree's modern name is *Cordia macrophylla*. or, commonly, fishleaf or man-jack.

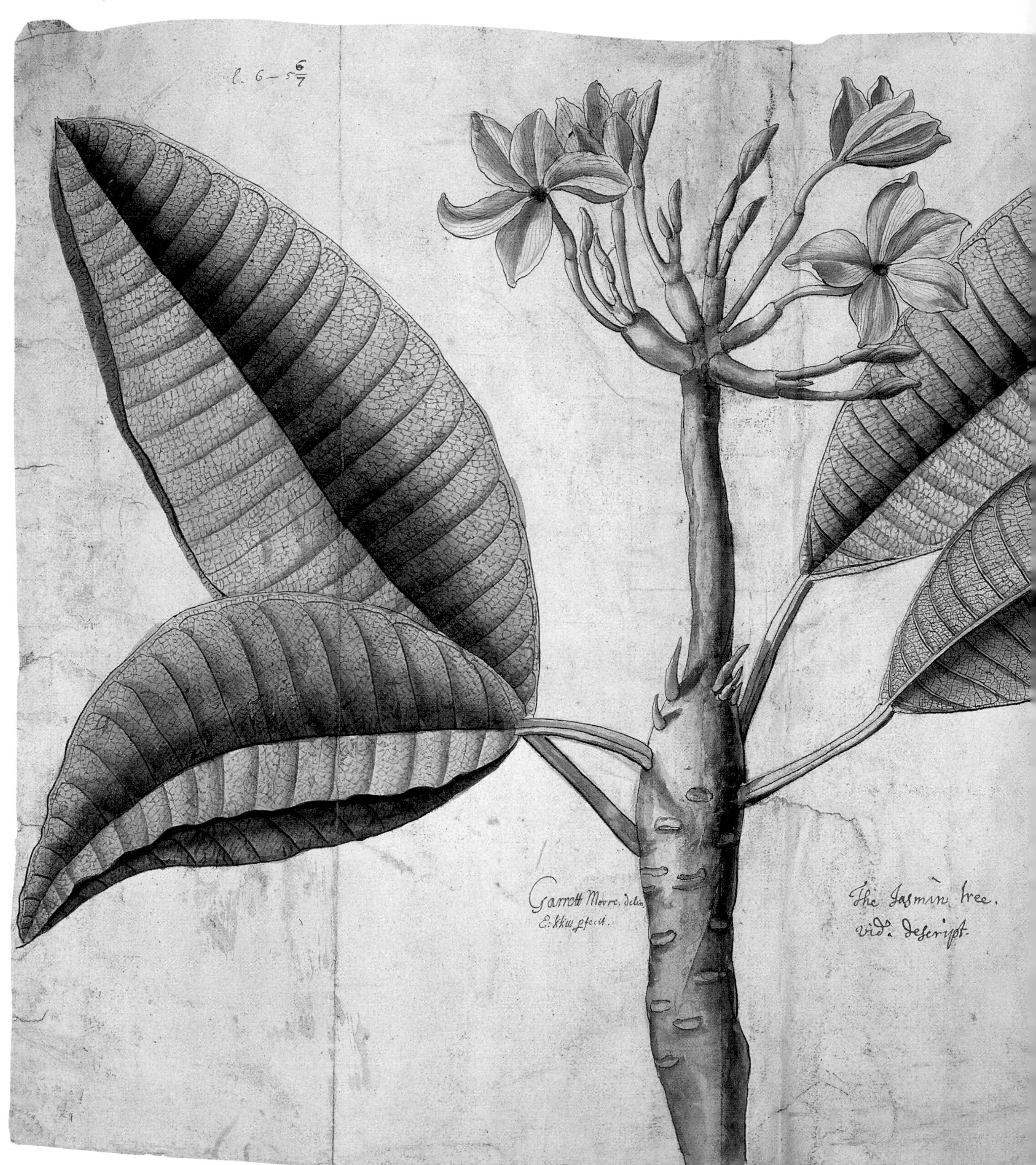
Garrett Moore, delin,
The Jasmin tree.
vid. descript.

***The 'Jasmin-Tree'** (left)* or *Nerium arboreum, folio maximo obtusiore, flore incarnato* (now *Plumeria rubra*). Sloane observed that it was about the size of an apple tree, and that it had 'pleasant Flowers both for Colour and Smell'. It was grown as an ornamental tree in the gardens of Jamaica, Barbados and the 'Caribe Islands', but had practical uses, too. 'The Tree yields a Milk of a burning Nature,' Sloane wrote, 'and yet the Indians say that taken ... [as] two Scruples twenty four Grains, it purges very easily the Phlegmatic, Cachectical Humours of those in the French Pox or Dropsy, especially if they come from a cold Cause.' A specimen *(below)*, *Tradescantia zanonia,* from Volume 4 of the herbarium.

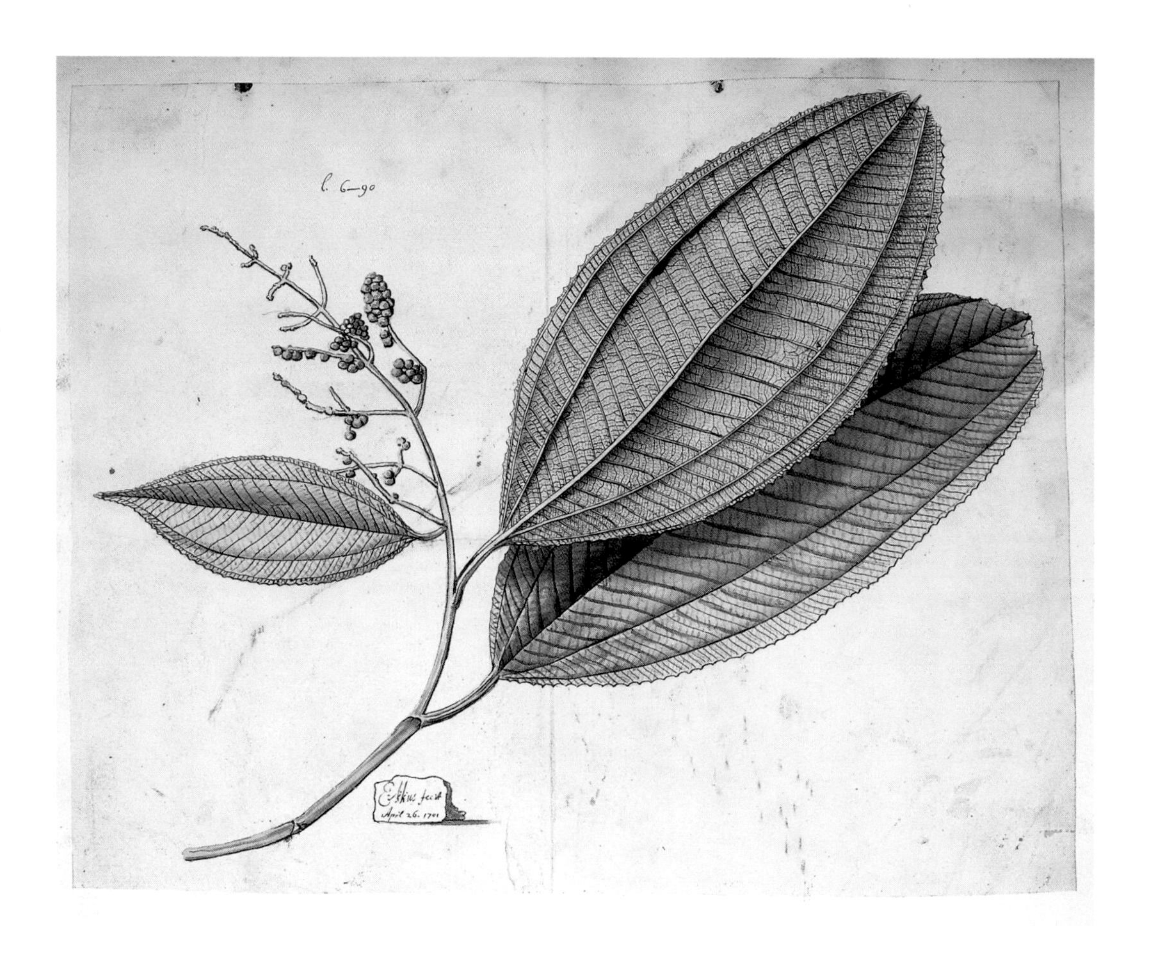

A specimen *(opposite)* ***and a Kickius drawing*** *(above)* of a tree described by Sloane as *Grossulariae fructu arbor maxima non spinosa, Malabathri folio maximo inodoro, flore racemoso albo* (now simply *Miconia elata*). The drawing is dated 'Aprill 26 1701'. Sloane remarked on its trunk 'as thick as one's Thigh, cover'd with a russet colour'd, almost smooth Bark, very streight and twenty Foot {6 metres} high ...' Sloane concluded: 'It grew on the Inland Mountainous Woods, as about Mount Diablo, on the red Hills, near and beyond Colonel Cope's Plantation ... and in Barbados.'

Bromelia P.

Caraguata-acanga Pis. cat.

The plant bearing these fruits *(opposite)* was identified by Sloane as *Caraguata-acanga* (now *Bromelia pinguin*). He remarked: 'It quenches Thirst extremely, and on the landing of the English forces on Hispaniola, in their want of water, was thought to save many Lives by that its quality.' It had its drawbacks, however: 'The Fruit is very acceptable by reason of its grateful acidity, but it not only sets the Teeth speedily on edge, but likewise brings the Skin off of the Roof of the Mouth and Tongue.' Taken with a spoonful of sugar, it was used to cure worms, thrush and mouth ulcers, and was also thought beneficial in cases of fever and as a diuretic. It also had the power to induce miscarriage 'of which Whores being not ignorant make frequent use of it to make away their Children.'

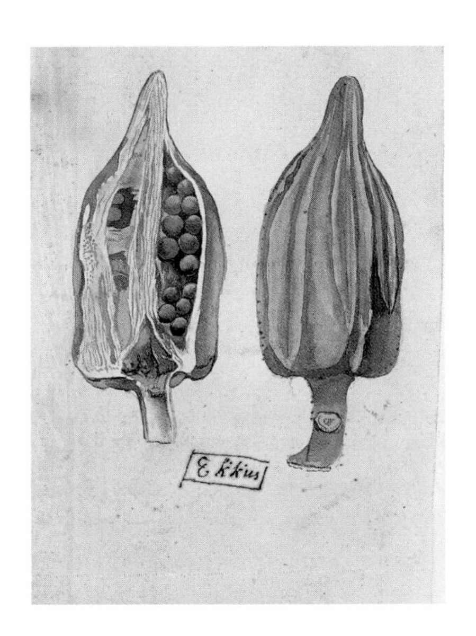

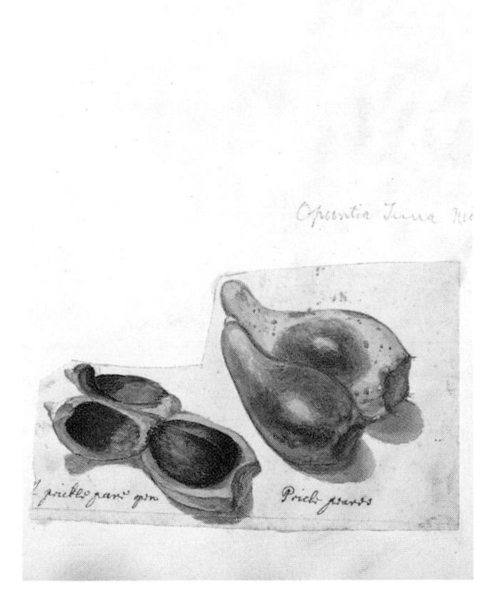

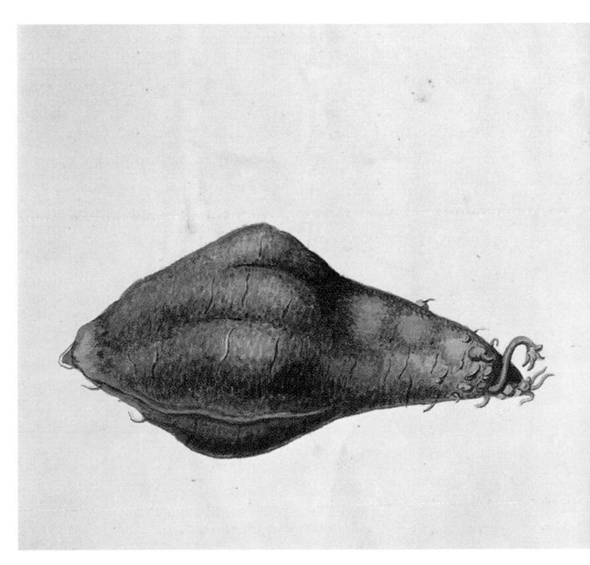

The pods of okra *(above left),* ladies' fingers, or bhindi, which has the modern botanical name *Abelmoschus esculentus.* The fruit of the 'Prickly Pear-Tree' *(above centre)* – which has the modern name *Opuntia spinosissina* – had juice with the ability to dye 'Linen ... as well as Mouth and Hands, or whatever it touches ...' Perhaps the most familiar of these plants is the sweet potato *(above right),* or *Ipomoea batatas.* Sloane referred to it as 'Spanish Patatas', explaining that: 'They are boild or roasted under the Ashes, and thought extraordinary good and nourishing Food, and because of their speedy attaining their due growth and perfection, they are believed to be the most profitable sort of Root for ordinary Provision.'

Surveying Ceylon 1672–1757

Paul Hermann, Johan Gideon Loten & Pieter de Bevere

Plantains sprout from the fruiting head of *Musa x paradisiaca (opposite)* in Paul Hermann's volume of drawings. These heads can weigh from 18 to 27 kilograms (40 to 60 pounds). The fruits are only edible once cooked.

The Natural History Museum's collections in London contain the works of hundreds of unsung heroes, men and women whose efforts have been largely overlooked though they may have been significant in the development of the ideas of more celebrated scientists. Two examples represented in the Museum originated in Ceylon, now Sri Lanka, long before the British involvement with the island; they also illustrate how long-lived institutions like the Museum sometimes acquire material many years after it was collected or produced – and sometimes from unexpected original sources.

The Dutch took possession of Ceylon from the Portuguese in 1658 and held it for 140 years until it became a British possession in 1798. During this period the island was effectively managed by the Dutch East Indies Company under a series of Governors. One might have expected that scientifically valuable or interesting material found in Ceylon during this time would have found its way to one or other of the excellent collections in museums and universities in the Netherlands. But in 1925 a bookseller in the Hague, Messrs Martinus Nijhoff, offered for sale a collection of illustrations of Ceylonese plants and animals and a small amount of manuscript material assembled by Johan (or Joan) Gideon Loten who had been Governor of Ceylon from 1752 to 1757. On the recommendation of Norman Kinnear, then an assistant in the Zoology Department of The Natural History Museum in London and later to be knighted as the Museum's Director, the Museum purchased the collection for £75.

Loten was born in 1710 in the municipality of St Maartensdijk in the Utrecht province of the Netherlands and worked for the Dutch East Indies Company from 1731 until 1757. During his first 20 years with the company he served in increasingly important roles in Batavia (now Jakarta), Samarang, and Macassar in Sulawesi, marrying South African-born Anna Henrietta van Beaumont in 1733. Finally, in 1752 he was appointed Governor of Ceylon and he and his wife moved to Colombo. Loten's five-year governorship was not an easy one. He had arrived on the island at a time of great political unrest, characterised by almost continual internal disturbances amongst the native Sinhalese. Under his successor this instability escalated to a costly and full-scale war between the Sinhalese and Dutch, but even in Loten's time,

Hermann's original book of illustrations *(opposite),* which accompanied his herbarium. These pages show a leaf, flower and seedhead of *Nelumbo nucifera*, the sacred lotus or Egyptian bean. Traditionally, the flower was sacred in China, Tibet and India, where it was known by its Sanskrit name, *padma*. The specific name *nucifera* means 'nut-bearing', and if embedded in river mud, the 'nuts' or fruits of *N. nucifera* can remain viable for centuries.

the situation was precarious. There was also the constant possibility of external interference, particularly from the French or British. Furthermore, to add to his anxieties, his wife died in 1755.

Despite all this, Loten maintained a passionate interest in science, and particularly natural history, which he had pursued during his earlier overseas postings. In Colombo he made a collection of the local animals and plants and employed a talented local artist to make paintings of them. Little is known about this artist and different sources disagree even about his name. His family name, de Beveren or de Bevere, is a well-known one in Holland and came from his paternal grandfather, Willem Hendrik de Beveren, a Dutch military officer who had a son with a local Sinhalese woman. This son, the artist's father Willemsz de Bevere, was born in Batavia about 1700, married and moved to Colombo as an assistant in the East Indies Company. It was here that the artist was born around 1733. Somewhat confusingly, he is variously referred to as Willem Hendrik, the same as his grandfather and common practice at that time, or as Pieter Cornelis. Almost nothing is known of his youth, education or earlier employment, nor about his later life other than that he moved to Batavia with Loten in 1757, when he was in his early twenties, and had died by 1781.

But his excellent drawings and paintings survive, produced mostly in Ceylon but a few in Batavia. The 154 in The Natural History Museum's Loten collection include 98 of birds, five of mammals, seven of fish, 17 of assorted invertebrates and 16 of plants. They have never been published in their own right, but they were well known among 18th-century natural historians. This was probably because, after a year or so in Batavia, Loten retired, first to the Netherlands and then to London where he lived in Fulham from 1759 to 1765 and married an English woman, Letitia Cotes. He spoke and wrote English extremely well and was known and respected by the capital's scientific community, so much so that he was elected a Fellow of the Royal Society in 1760. During this period, and after he later returned to the Netherlands, he made the de Bevere paintings available for use in a number of important and influential publications of the time. They were used, for instance, by George Edwards, librarian of the Royal College of Physicians, in his *Gleanings of Natural History* published between 1758 and 1764. They were also used by J. R. Forster, naturalist on Captain Cook's

second Pacific voyage in his *Indische Zoologie* published in 1781, and by Thomas Pennant in his *Indian Zoology*, published in 1769. For Pennant's book Sydney Parkinson, Sir Joseph Banks' draughtsman on Cook's first voyage, copied the de Bevere paintings. This was more or less the last use of the de Bevere drawings in Loten's lifetime as Loten died in Utrecht the same year that Pennant's book was published.

Loten had bequeathed the de Bevere paintings to the Society of Sciences in Haarlem where they stayed until 1866 when the Society's collections were for some reason dispersed. They then seemed to disappear without trace for almost 20 years until, in 1883, they were offered for sale by Messrs Nijhoff in the Hague, the same firm of booksellers who later sold them to the Museum. This time, however, they were sold for 300 florins to a Mr P. J. van Houten who did considerable research into their background and subsequently became President of the Commission of the Colonial Museum at Haarlem. Why they did not end up in this museum or in some other appropriate Dutch institution is a mystery. Perhaps no one would pay Mr Houten what he considered was a reasonable price for them. At any rate, Mr Houten or his executors apparently sold them back to Messrs Nijhoff some time in the early 1920s and they came to their final resting place in The Natural History Museum in London, no more than a kilometre or two from where Johan Loten would have kept them during his residence in Fulham some two-and-a-half centuries previously.

Almost exactly a century before the Loten collection was acquired, the British Museum received another much earlier, but equally important, collection of Ceylonese material. This one arrived in 1827 as part of the massive collections of Sir Joseph Banks, but it dated from the very early days of Dutch control of Ceylon. In the mid- to late-17th century, the island was an important strategic stronghold for the Dutch, but it was also prized as an important source of natural products – particularly cinnamon which was greatly valued in Europe for its reported effectiveness in removing wind from the bowels. By 1664, the annual exports of cinnamon to Amsterdam had increased from about 113,400 kilograms (250,000 pounds), when the Dutch East Indies Company first took control of the plantations, to 680,400 kilograms (1.5 million pounds). Several years after this peak in the cinnamon trade, in 1672, a young doctor, Paul Hermann, began

a five-year assignment as Chief Medical Officer to the Dutch East Indies Company on the island and showed a great interest, not only in cinnamon production, but in all aspects of the island's botany.

Born in 1646, Hermann had not long completed his medical studies when he went to Ceylon. He was clearly more interested in botany than in the closely allied subject of medicine and would have seen this as a superb opportunity to collect plants from a region that had barely been explored by Europeans. Although the island was nominally a Dutch possession, control was limited to the coastal regions, as had been that of the Portuguese before them. The whole of the interior was still held by the indigenous Sinhalese ruler, Emperor Raja Singha, so that opportunities for travel into the untouched forest areas were severely limited.

The herbarium of preserved plants that Hermann put together during his stay reflects this, being largely representative of the flora in the general area of Colombo where he was based and including plants from local gardens. As a result, about 50 of the several hundreds of plants that Hermann collected were exotics, not native to Ceylon at all. Most of these are native European species, presumably taken out by Portuguese and Dutch colonials to grow in their gardens and remind them of home. A dozen or so, including the custard apple, guava, cashew nut, capsicum and cotton, are from the Americas and are a graphic example of how quickly plants were transported from one relatively newly-explored locality to another. But despite these 'contaminants', and the rather restricted coverage of the Ceylonese flora that Hermann's collection represents, the four volumes of dried plants, and the accompanying volume of original drawings, amounted to an extremely important collection for the botanists of the day.

Hermann himself had very little opportunity to publicise the collection in his lifetime. Following his Ceylon sojourn, at the age of 32, he was appointed to the Chair of Botany at Leiden in 1679, a post he held until his early death in 1695. But apart from a brief catalogue of his collection that appeared in 1687, he appears to have published nothing on the botany of Ceylon. After his death his widow sent the herbarium and her husband's manuscript notes to William Sherard, Professor of Botany at Oxford, presumably hoping to make a little money from any resulting publication. Sherard edited

Although, as a doctor, Hermann's main interest was in the botany of Ceylon and its possible medicinal applications, he could not resist recording some of the animals of the island, such as this slender loris *(opposite)*.

the notes to produce a 71-page catalogue of the plants under their Sinhalese names that was published anonymously in 1717 under the title *Musaeum Zeylanicum*. No one seems to have taken much notice of it, or of the associated collection until it turned up almost 30 years later, this time with August Günther, Apothecary-Royal in Copenhagen. Knowing of the interest of the by-now celebrated Carl Linnaeus in examining plants from exotic localities, Günther sent the whole collection to Uppsala, in Sweden, and Linnaeus started work on it immediately. Two years later, in 1747, he published *Flora Zeylanica*, listing 657 plants from Hermann's collection, 429 of them already assigned to one or other of his 'genera' and cross-referenced to his own numbers and notes on Hermann's herbarium sheets. This elevated Hermann's material to a very important position in the history of botany.

Linnaeus, the son of a Lutheran priest, was born in 1707 and studied medicine at the University of Uppsala, against the wishes of his parents who had hoped that, like his father, he would enter the church. During his twenties he made a series of botanising journeys in north-western Europe, to Lapland, Germany, the Netherlands and England, where he met Hans Sloane. Then, after a short period practising medicine in Stockholm, in 1741 he was appointed to the Chair of Botany at his old university. The *Flora Zeylanica* was just one of a whole series of similar works which he published, describing collections of plants either obtained from relatively restricted areas, like this one, or accumulated from a wide range of sources. They represent important stages in the development of his ideas which led ultimately to his gift to the world of natural history – the binomial system.

Binomial means simply 'two names' and refers to the fact that, ever since Linnaeus introduced the system in the mid-18th century, botanists and zoologists all over the world have given each plant or animal species a two-part Latin name. The first part, the generic name, is common to all organisms considered to be sufficiently similar to belong to the same group, or 'genus'. Together with the generic name, the second part – the specific name – provides a unique combination which applies to that species alone. Thus a human is *Homo sapiens,* a lion is *Panthera leo* and a daisy is *Bellis perennis.* Consequently, any scientist knows, or should know, exactly what another scientist means by the use of

The ball-lobed dead-leaf mantis *(opposite), Gongylus gongylodes*, painted by Pieter de Bevere for Governor Loten. The 'dead leaf' epithet refers to the insect's ingenious form of disguise. Voracious predators, mantids are powerful mimics of such natural plant forms as leaves or even flowers, which enables them to wait undetected until an insect victim is close enough to be seized by a lightning-fast strike of the mantid's front legs. In this species, the knee joints of the hind legs and the second section of the front legs are greatly enlarged; hence the generic name, *Gongylus*, which means 'ball' in Greek.

such a 'binomial'. This was an enormously significant advance because there had previously been no universally agreed naming system. Instead, the scientific names used tended to be complete Latin phrases describing the characteristics of the species but resulting in appalling confusion and misunderstanding. Since Linnaeus had not yet devised his simple binomial naming system the plants in *Flora Zeylanica* are not named in the modern sense. But when he published *Species Plantarum* in 1753 all the species were 'properly' named and Linnaeus carefully referred them back to *Flora Zeylanica* so that the names can be linked directly to Hermann's herbarium. The 'type' concept, by which botanical and zoological names are associated with specific voucher specimens of the plant or animal, was unknown in Linnaeus's day and was a much later invention. Nevertheless, in retrospect the Hermann specimens are 'types' of the names which Linnaeus created for them and are therefore treasured by The Natural History Museum, where they now reside together with the accompanying volumes of illustrations. But their route from Linnaeus to South Kensington is largely thanks to a well-established 18th-century network of collectors.

Having finished his scientific examination of the material, Linnaeus sent the Hermann collection back to Günther who gave or sold it to Count A. G. Moltke, also presumably in Copenhagen, or at least in Denmark. When Moltke died the collection was bought by Professor Treschow, again in Copenhagen, so that by now the Hermann material had spent several decades in Denmark. But Treschow ultimately sold it for £75 to that great accumulator of collections, Sir Joseph Banks, whose vast hoard of treasures finally came to the British Museum in 1827 and then to The Natural History Museum in London in 1881.

A palm plantation *(left)*, in a drawing from the Hermann collection, showing various stages in the production of the alcoholic toddy made from the palm's sap. On the far left and centre right, two men can be seen climbing the palms to extract the sap. On the far right the raw sap is fermented or distilled in a vat or still. In the centre, a group of Europeans are smoking and drinking, while on the far left a man appears to be vomiting – perhaps from overindulging in the plantation's product. As well as being an alcoholic drink, toddy was traditionally used in India in place of yeast, as a leavening agent, in the same way that grape must was used in other parts of the world.

A single leaf *(above)* ***and a whole plant*** *(opposite)* of *Nymphaea pubescens*, the night lotus, as it appears in Hermann's volume of drawings. The common name 'lotus' is used to describe plants from a number of different genera – this species belongs to one of the genera of the water lilies. Several of the so-called 'lotuses' have an important place in eastern religion. With its white flowers, *N. pubescens* is sacred to Shiva, the Hindu dancing god and one of the three deities in the supreme triad of Shiva, Brahma and Vishnu.

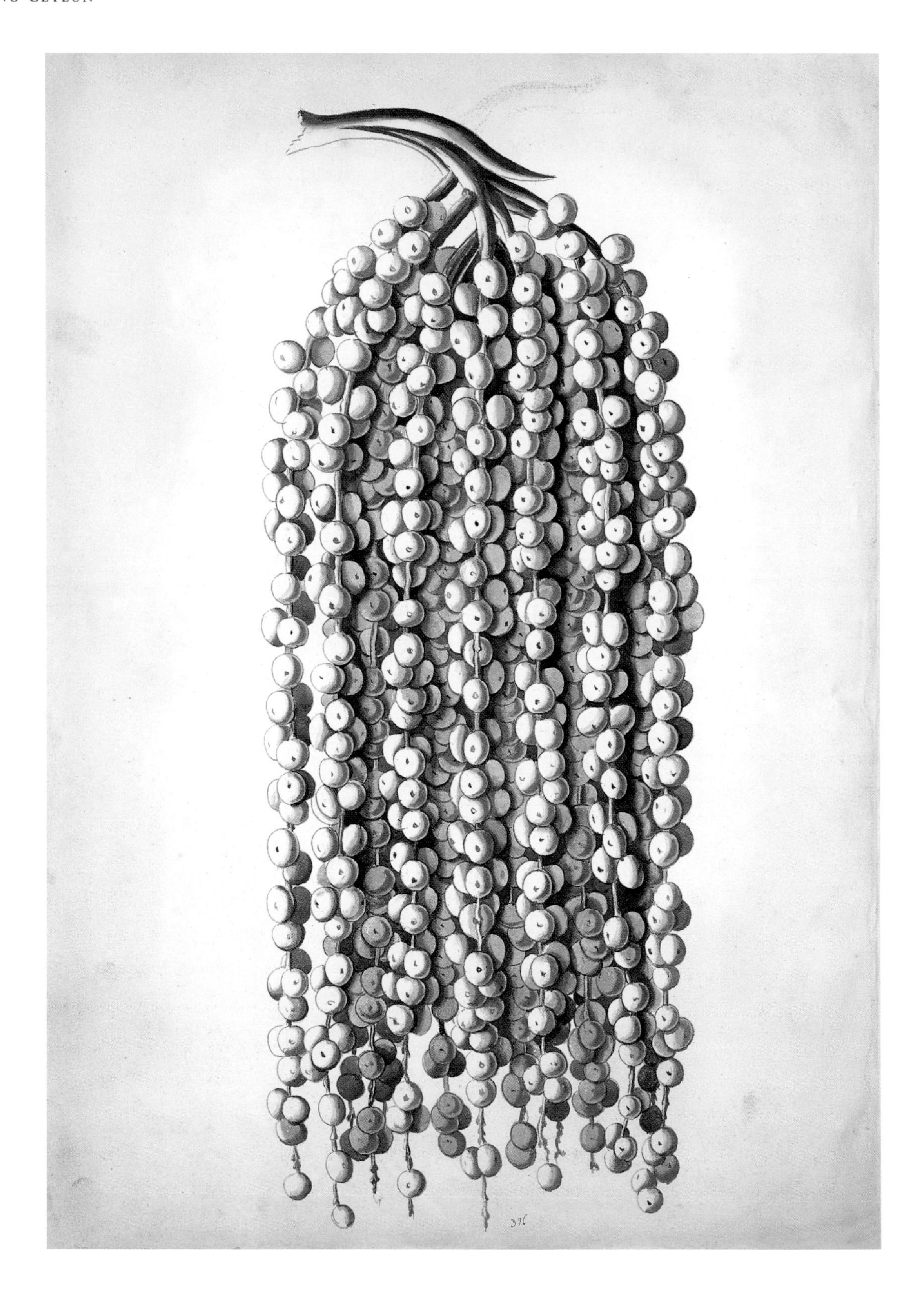

The spadix or dense flower tassel *(above)* of *Caryota urens*, the toddy palm. It is from the incised spathes – the modified leaves at the base of the spadices – that sap is taken and fermented to produce toddy. A single inflorescence can yield 7 to 14 litres (1½ to 3 gallons) a day, but if more than one is tapped, as much as 27 litres (6 gallons) can be obtained. Although the globular red or yellow fruits of a fruiting inflorescence *(opposite)* contain an irritant, the seeds are nevertheless dispersed by animals. In the wild, the tree grows in clearings and on the edges of the rainforest subcanopy; it is also widely grown in cultivated gardens.

396

Caryota urens, ***as illustrated in Hermann's collection*** *(opposite),* or the toddy palm, also a source of sago, kitul and jaggery. Its young leaves are edible, its pith is used to make the starchy foodstuff known as sago, and its sap is evaporated to produce jaggery (a type of coarse sugar) or fermented to make toddy. Kitul, a type of fibre, is derived from its leaf-stalks and used in the making of baskets and ropes. Traditionally, the tree had medicinal uses, too: the oil of a young palm was applied in paste or poultice form to treat snake bites. Other plants in Hermann's volume of drawings are not as well known *(below, left & right).* When Carl Linnaeus examined these drawings in around 1745, he was unable to identify them, perhaps because they are too highly stylised. They still remain unnamed.

Birds feature strongly in the Loten collection. Pieter de Bevere's painting of the black-rumped flameback, or the lesser flame-backed woodpecker, *Dinopium benghalense (above)*, included just enough additional information to suggest the subject's habitat and behaviour, without sacrificing the overall clarity of the image. In another of his illustrations *(opposite)*, a paradise flycatcher, *Terpsiphone paradisi*, sits below a purple-rumped sunbird, *Nectarinia zeylonica*. The flycatcher feeds on insects which it catches on the wing.

One of the 98 bird illustrations *(above)* by de Bevere now owned by The Natural History Museum shows a dead kingfisher lying across a tree stump. The species is *Pelargopsis capensis,* the stork-billed kingfisher. This is one of the few illustrations of the period that shows the artist's subject in its true state, as nearly all were painted after they had been caught and killed. All the more remarkable, then, is the way in which de Bevere imbues the rest of his animal subjects with such life, as in the depiction of *Psittacula eupatria,* an Alexandrine parakeet *(opposite),* peering down from a branch.

The bird that to Europeans epitomises exoticism and opulence – the peacock, *Pavo cristatus (right)*. The species is native to Sri Lanka and India. The peacock feathers, which the male displays to attract a mate, are not in fact his tail feathers but the enormously elongated and erectile tail coverts, feathers that grow from the lower part of the back.

Perched on a flowering branch, this is *Otus bakkamoena (opposite),* in a finely worked illustration by de Bevere. It is commonly known as the collared scops owl or the little horn owl – because of its longish ear tufts that protrude like horns above its head. The caterpillars *(above)* of one of the emperor moths, *Antheraea sp.*, crawl over a leafy stem, while below hangs a cocoon of the same species. Emperor moths belong to the family *Saturniidae*, which produce large amounts of silk thread when forming their cocoons.

The atlas moth, **Attacus atlas** *(below left),* as painted by de Bevere. It is native to India and South East Asia and belongs to the *Saturniidae* family, in which the males are characterized by comb-like antennae. These sensitive scent monitors serve an essential function for the survival of the species – enabling the male to detect the presence of a female atlas moth at some distance away, by picking up the sex pheromone she emits. The fungus *Phallus indusiatus (below right)* is fairly common in tropical regions of Asia. Because of its shape and smell, it is commonly called the veiled stinkhorn.

This unusual view of the underside of the atlas moth *(opposite)* is another in the series of studies by de Bevere of this remarkable creature, which possesses the largest wingspan of any flying insect – measuring up to 30 centimetres (12 inches) across. Like all moths and butterflies, the atlas moth has membranous wings covered with rows of minute scales, set close together like roof tiles or the pieces in a mosaic to create the patterns and colours we see. The Greek name *Lepidoptera*, the order to which all butterflies and moths belong, translates as 'scale wings'.

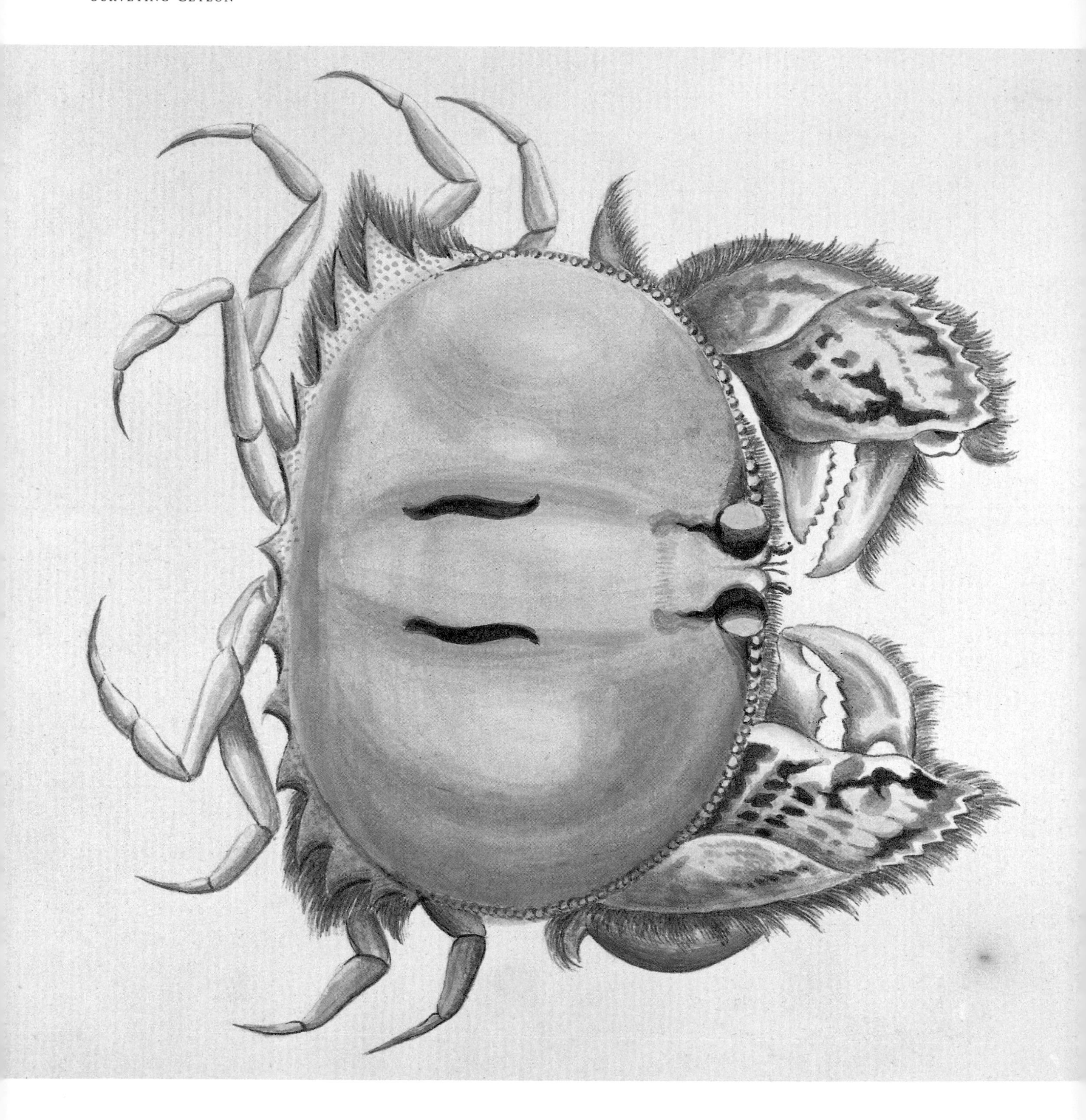

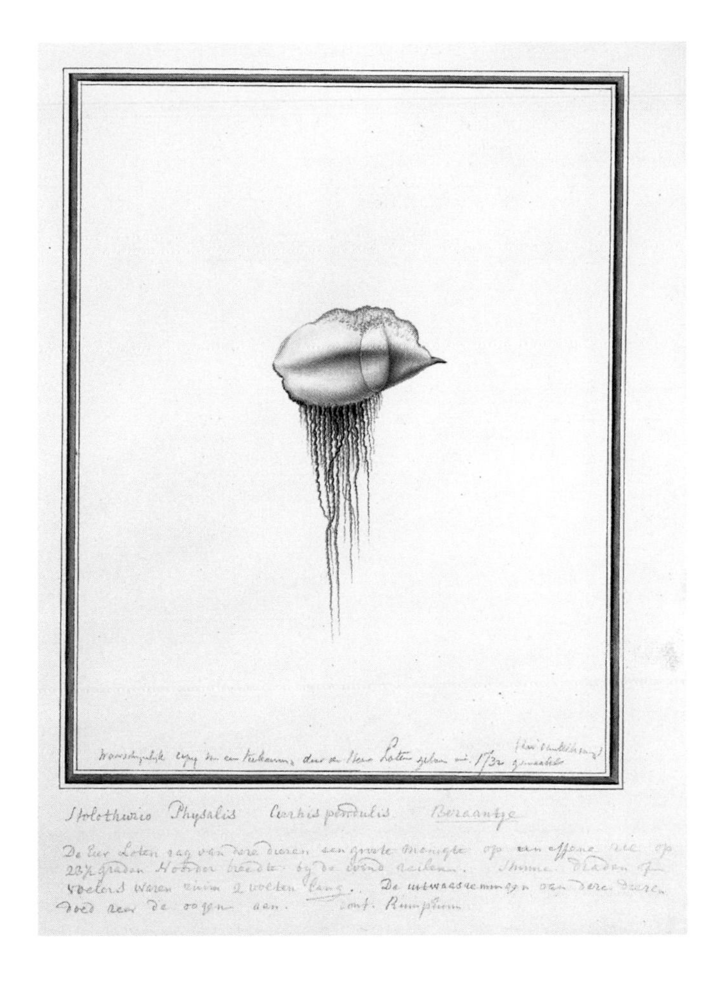

A number of the de Bevere paintings are of fish and sea creatures, including *Calappa philargius (opposite),* commonly called the stone crab or box crab – for the box shape it takes on when it folds its pincers tightly up against its carapace – and a Portuguese man-of-war, or *Physalia physalis (above).* The Portuguese man-of-war is not a single organism but a floating colony of individuals clustered below a gas-filled sac and protected by a drifting curtain of stinging tentacles.

The Sri Lankan giant squirrel or grizzled giant squirrel, *Ratufa macrura (above),* is native to Sri Lanka, but is also found in Tamil Nadu, a region in the extreme south of peninsular India. Like other squirrel species, the animal has long, sharp claws which give it a good grip when clambering around trees, while its excessively long tail acts as a balancing aid when running and leaping. This illustration is one of only five mammal paintings by de Bevere now in the possession of The Natural History Museum, including his rendering of *Semnopithecus vetulus (opposite),* commonly known as the purple-faced leaf monkey.

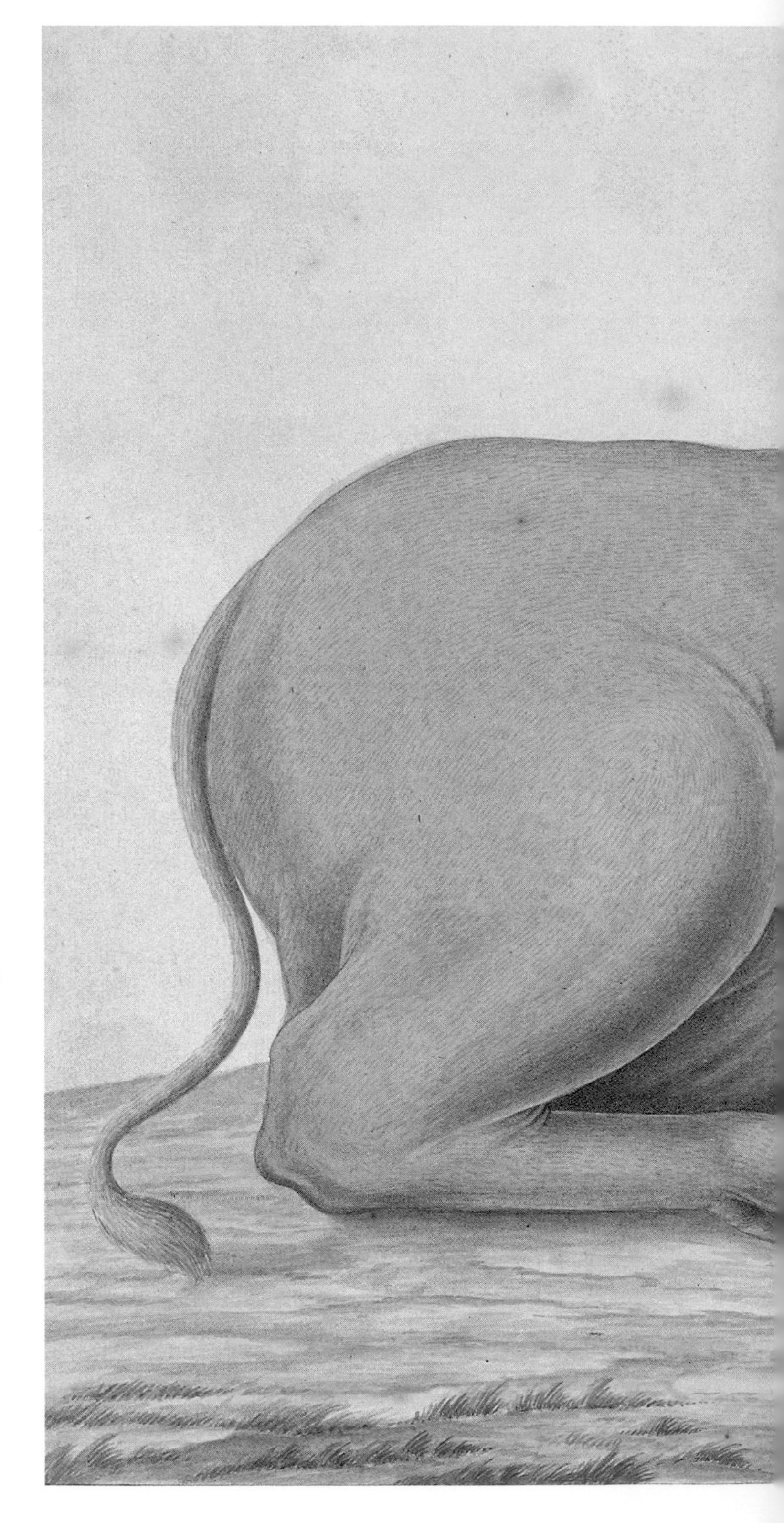

The babirusa *(right), Babyrousa babyrussa.* The male of the species has a double pair of curving, upturned tusks, reminiscent of the horns of certain deer, hence the species' common name of 'deer hog' (*babi* is Malay for 'hog' and *rusa* the Malay term for 'deer'). *B. babyrussa* is found on the island of Sulawesi in eastern Indonesia, and on neighbouring Togian and Sula islands; it also inhabits Buru island, where it may have been introduced. This would have been one of the animals that de Bevere saw and drew when he moved to Batavia (now Jakarta) with Loten in 1757.

The Indian chevrotain or mouse deer *(overleaf), Moschiola meminna.* Although superficially it has a deer-like appearance, it is not a true deer but belongs to the *Tragulidae* family. Etymologically, both its family name and common name relate it to the goat: *Tragulidae* comes from the Greek *tragos*, or 'goat', while the common name chevrotain derives from *chèvre*, the French word for 'goat'. The species is native to Sri Lanka and peninsular India.

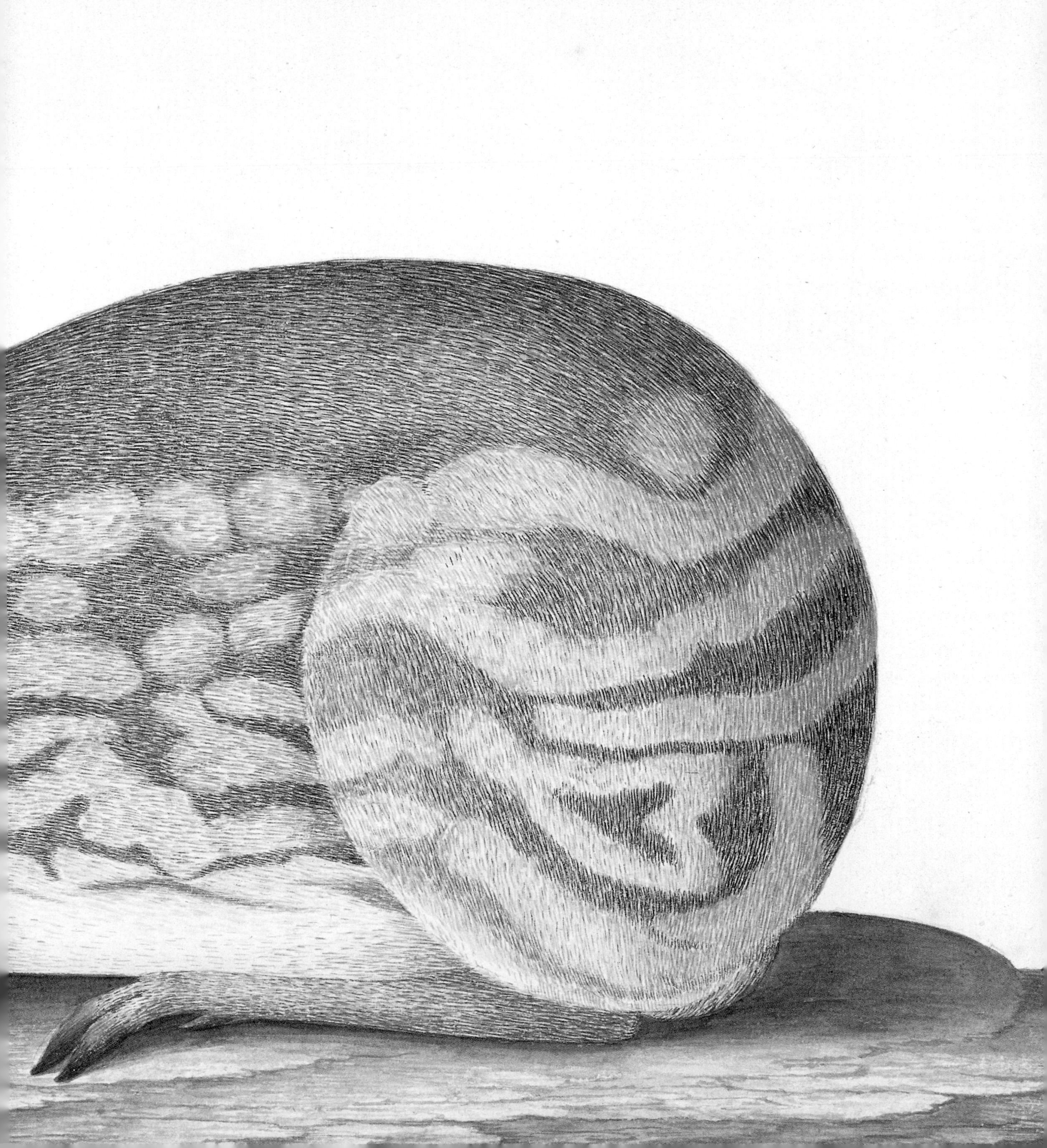

Sojourn to Surinam 1699–1701

Maria Sibylla Merian

During her two years in Surinam, Merian beautifully rendered insect and butterfly life cycles. In one example *(opposite),* the caterpillars of the emperor moth, *Arsenura armida,* feast on the leaves of the coral tree, *Erythrina fusca.*

Amongst Hans Sloane's collection of original natural history drawings was a set of superb watercolour paintings which had been used to illustrate a book on the butterflies and moths of the Dutch colony of Surinam, published in 1705. The paintings, and the book, were the work of a remarkable woman, Maria Sibylla Merian, and they earned her both the admiration of her contemporaries, including Sloane, and the respect of subsequent generations of entomologists and students of natural history illustration. This is not only due to the undoubted artistic and scientific merit of the paintings and the accompanying text, but because in order to produce them Merian undertook an extremely uncomfortable and dangerous journey, under circumstances that would have been considered quite radical for late 17th-century times – a European woman of 52 travelling to the New World alone with her daughter and without the protection of a man.

Maria Merian was born in Frankfurt am Main in 1647, the daughter of a well-known publisher and engraver, Matthäus Merian the Elder. Her father died when she was only three years old but the following year her mother, Johanna Sibylla, married still-life painter Jacob Marel, so Maria was exposed to the artistic world throughout her formative years. In 1665 she herself married an artist, Johann Andreas Graff, and after the family moved to Nuremberg in 1670 she worked as a flower painter and engraver, publishing three volumes of flower engravings between 1675 and 1680. From an early age, however, she had been a keen student of entomology, and this subject gradually became the focus of her artistic endeavours. In 1679 she published a small book on the metamorphoses of European butterflies containing 50 copperplate engravings, uniquely displaying the full life cycle of the insects. This pioneering form of natural history illustration, and her fantastic eye for detail, earned her a reputation in both scientific and artistic circles, and greatly impressed Hans Sloane who purchased some of the original watercolours from which the engravings were made.

A second volume of her European butterflies was published in 1683, by which time Merian, her husband and two young daughters had moved back to Frankfurt. But Maria's marriage was an unhappy one, and in 1685 she left Graff to take her daughters to join a Protestant spiritual regeneration community, the Labadists, based at the castle of Waltha near

Leeuwarden in the Netherlands province of Friesland. As was fashionable for rich Dutch burghers of the time, the owner of Waltha Castle, Cornelis van Aersen van Sommelsdijk, had a private collection of natural objects such as shells, bird skins, minerals and butterflies from the Netherlands' overseas colonies, including insects from Surinam collected when van Sommelsdijk served as Governor. The butterflies in this collection were so exotic in size, form and colour compared with the European species with which Merian was familiar that her imagination was fired and she became determined to see them in their natural environment. Her chance came when she left the Labadist community in 1691 and moved to Amsterdam where she made the acquaintance of some of the most influential citizens in the capital, including the city's burgomaster Nicolaas Witsen. With the support of Witsen and other powerful friends, in 1699, she obtained a stipend from the Dutch Government and set sail in June, at the then relatively advanced age of 52, with her younger daughter on a two-month-long voyage to Paramaribo, Surinam.

The voyage would not have been a pleasant one in cramped, ill-ventilated and rat-infested accommodation, with monotonous and far-from-fresh food and very limited water. But Surinam would have been much worse. The climate, always hot, but alternately very wet or very dry, was extremely trying for those unused to it. Moreover, Surinam was a potentially violent country, for the inhumane attitude of the colonising planters and their overseers made slave revolts by no means uncommon. After two years, and despite the fascination of the luxuriant vegetation and teeming insect life, Merian decided to leave, and she and her daughter returned to Holland.

During her time in Surinam, Merian had accomplished a phenomenal amount of work, seeking out caterpillars, observing and illustrating them feeding on the host plant, watching them pupate and keeping the chrysalis until the adult emerged. Almost all of her observations were totally new and the resulting book, *Metamorphosis Insectorum Surinamensium,* prepared after her return to Amsterdam and including illustrations which she not only painted but also engraved herself, was widely acclaimed. A few of the 60 coloured engravings depict non-entomological subjects including frogs, toads, snakes, spiders and even an alligator, but the rest are devoted to butterflies and moths. Each plate shows one

or two species of caterpillar feeding on the host plant and the adults flying away to one side or the other. Although the composition is always artistically pleasing, accuracy is never sacrificed. Because the butterflies and moths were mostly unnamed at the time, only the plants are labelled. But such was the quality of Merian's observation that subsequent identification has rarely been a problem. Her work was published at a time when natural history accounts of exotic places were often illustrated by artists whose work was, at best, based on damaged and poorly preserved dead specimens, but more usually on imagination and garbled second-hand accounts. She set an extremely high standard for subsequent illustrators to aspire to; many of them failed.

The book, published in Dutch and Latin, sold very well at home and abroad, so well in fact that in 1726 it was published in French. But Merian did not live to see this edition, having died in Amsterdam in 1717 at the age of 69. Long before her death, however, Merian's original paintings, not only from the Surinam book but also from her earlier works, had begun to attract interest from collectors; her admirers included Peter the Great and some of her paintings are still in Leningrad. As an astute businesswoman, as well as an accomplished observer and artist, Merian recognised the commercial potential of this interest and she was not averse to maximising her profits. Accordingly, she produced more than one set of 'originals' of some of her work so that there are, for example, two sets of original watercolours for her Surinam book in London, one in the Sloane collection and one in the Royal Library at Windsor, England.

Merian's paintings, and the publications in which they were reproduced, justifiably continue to attract interest and admiration. But perhaps her most fitting and lasting accolade is that both her European and Surinam insect works were assiduously studied and referred to by the father of modern botanical and zoological nomenclature, Carl Linnaeus. Since the material on which Merian had based her paintings and descriptions had not been collected and preserved, Linnaeus was unable to examine the insects first hand. But he was so confident in the accuracy of her observations that when, in 1758, he published his great work, *Systema Naturae,* listing, describing and naming all 4,400 animal species in the world known to him at the time, he was happy to base several of them entirely on Merian's accounts.

A New Draught of SURRANAM upon the coast of Guianna Made and Sold by John Thornton Hydrographer at the signe of England Scotland and Ireland in the Minories London
Aeooreorum Regio
A Scale of Miles
1 2 3 4 5 6 7 8 9 10
Armadille hills
Kirton Creek
Delawar
Scotsmans
Marhart
Adomson
Smith
Vaudoo
Manton
Dodd and Moore
English
Parham hill
Truckeribo
Sir Martin Noelias plat.
Tomontibo
Sir Robbart Harley
Mapanny or Willughbys River
Pacurara
Thorne
Stabworthy
St. Walter Rawleighs upperhouse
St. Martins Iland
St. Wtr Rawleighs lower house
Kingsland
Toorarica
Morganam
Country
Hunter
Read
Onamoba
Onaraebe
Japo
Lt. Gen: Byam
Oranea
Otapo
Burning
Specklewood
Erackirack
Parker
Clayton
Cunningham
Walkins
Dullen
Tannett
Byam
Champion
Shuttelworth
Weshimbo
Gaust
Redford
Samson
Pilgrim
Humphres
Elsing
Darvill
Wason
Davis
Leekes
Mermo
Wena
Com-
Cashwini
Supemia
Tabeth
Cor: mi: co
Panama
Opegtenda
Wepo
Jubenter
Carebeba
Egerton House
Monareco
Windall
Nianton
Price
Cuit
Bride
Hollis
Forest
Funck tree
Wembelton
Redrock
Sermo Iland
Creek
Marvel
Strange
Westrap
Titus
Uparth
Noeharwharf
Paramaribo
Fort
Wyatt
Remer
Whood
Clark
Bootiman
Toygershole
Wiampebo
Commawina
good fishing here
Red Banck
Cot: ti co
Pramarno
Som Ocomoco Indeans dwell here
good trading heer for bread and drinck
Mud Creek
Surranam River
Leward bay
Byams Point
Flatts
Windward Passage
Breakers
Leward Passage

METAMORPHOSIS
INSECTORUM SURINAMENSIUM.
OFTE
VERANDERING
DER
SURINAAMSCHE
INSECTEN.
Waar in de Surinaamsche Rupsen en Wormen met alle des zelfs Veranderingen na het leven afgebeeld en beschreeven worden, zynde elk geplaatst op die Gewassen, Bloemen en Vruchten, daar sy op gevonden zyn; waar in ook de generatie der Kikvorschen, wonderbaare Padden, Hagedissen, Slangen, Spinnen en Mieren werden vertoond en beschreeven, alles in America na het leven en levensgroote geschildert en beschreeven.
DOOR
MARIA SIBYLLA MERIAN.

Tot AMSTERDAM,
Voor den Auteur, woonende in de Kerk-straat, tussen de Leydse en Spiegel-straat, over de Vergulde Arent, alwaar de selve ook gedrukt en afgezet te bekomen zyn; Als ook by GERARD VALCK, op den Dam in de wakende Hond.

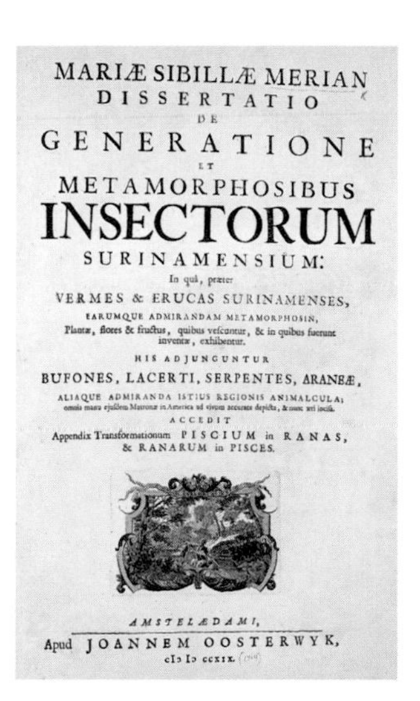

MARIÆ SIBILLÆ MERIAN
DISSERTATIO
DE
GENERATIONE
ET
METAMORPHOSIBUS
INSECTORUM
SURINAMENSIUM:
In quâ, præter
VERMES & ERUCAS SURINAMENSES,
EARUMQUE ADMIRANDAM METAMORPHOSIN,
Plantæ, flores & fructus, quibus vescuntur, & in quibus fuerunt inventæ, exhibentur.
HIS ADJUNGUNTUR
BUFONES, LACERTI, SERPENTES, ARANEÆ,
ALIAQUE ADMIRANDA ISTIUS REGIONIS ANIMALCULA;
omnia manu ejusdem Matronæ in America ad vivum accurate depicta, & nunc æri incisa.
ACCEDIT
Appendix Transformationum PISCIUM in RANAS, & RANARUM in PISCES.

AMSTELÆDAMI,
Apud JOANNEM OOSTERWYK,
CIƆ IƆ CCXIX.

The title pages of the 1705 edition *(above left)* and 1719 edition *(above right)* of *Metamorphosis Insectorum Surinamensium*, Merian's major work. The paintings for the book were completed during her two-year stay in the Surinamese capital, Paramaribo, marked in the lower right-hand corner of an early English map of the northern corner of Surinam *(opposite)*. That it was published at all is a tribute to her determination: 'Now that I am returned to Holland and several nature-lovers have seen my drawings, they pressured me eagerly to have them printed,' Merian wrote. 'They were of the opinion that this was the first and most unusual work ever painted in America. However, the costs involved in carrying out such a project made me hesitant at first; but finally I resolved to do it.'

***A branch of* Citrus aurantium** *(overleaf, left)*, or the Seville orange tree. The caterpillar, chrysalis and adult are *Rothschildia aurota*, in which Merian saw commercial potential: 'One finds these caterpillars frequently,' she wrote. 'They become so fat that they roll about; they appear three times a year; they spin a strong thread which made me think it might make good silk; I therefore gathered some and sent it to Holland where it was considered good, so if someone were to take the trouble to gather these caterpillars it would provide good silk and yield a good profit.' The water hyacinth *(overleaf, right)*, *Eichhornia crassipes*, commands the centre of this illustration, with the flying insect *Lethocercus grandis* overhead and its nymph below right. At the base of the plant is the spawn of *Phrynohyas venulosa*, while the tadpoles, juvenile frogs and adult frog are shown just above.

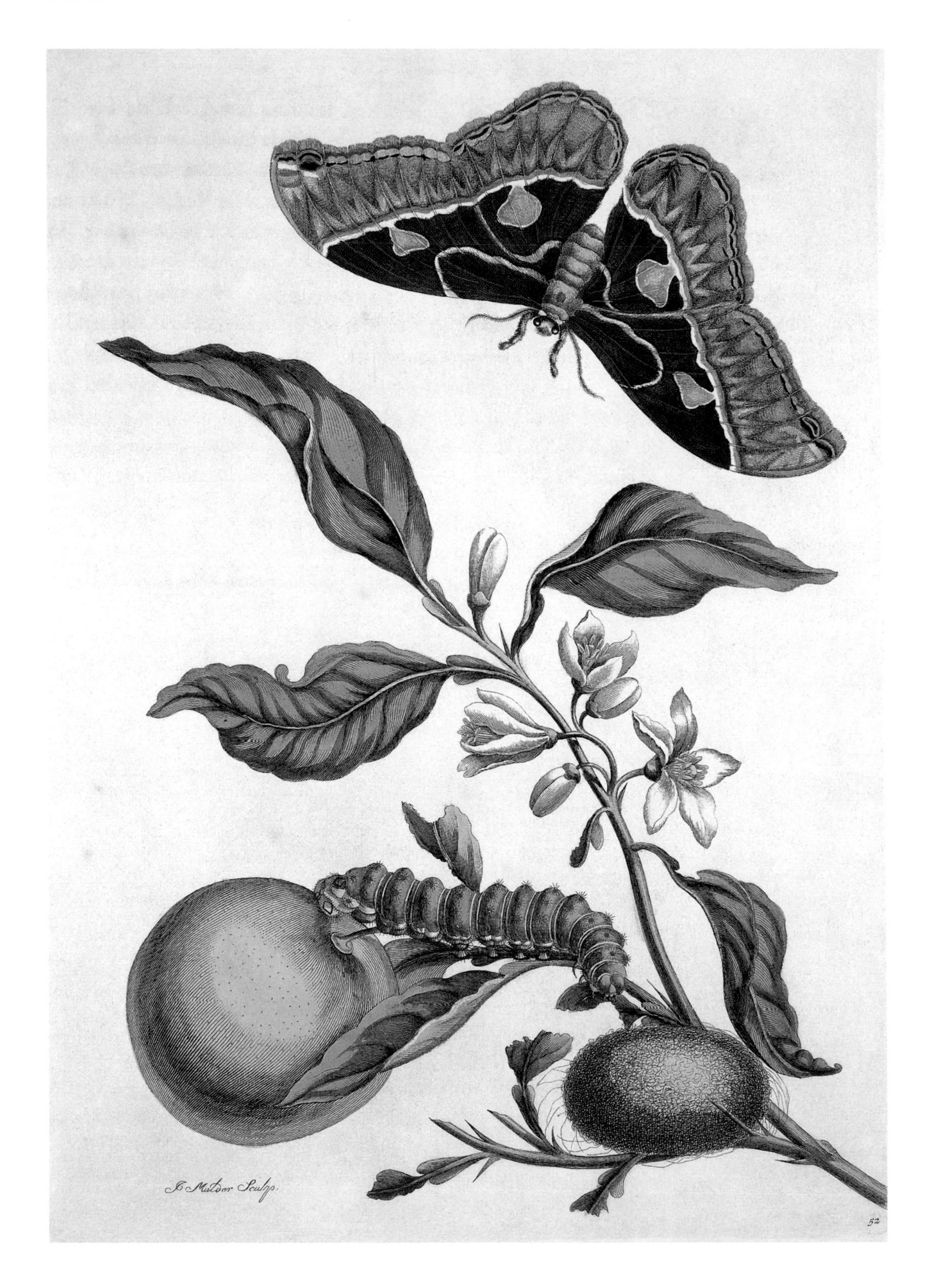
J. Mulder Sculp.
52

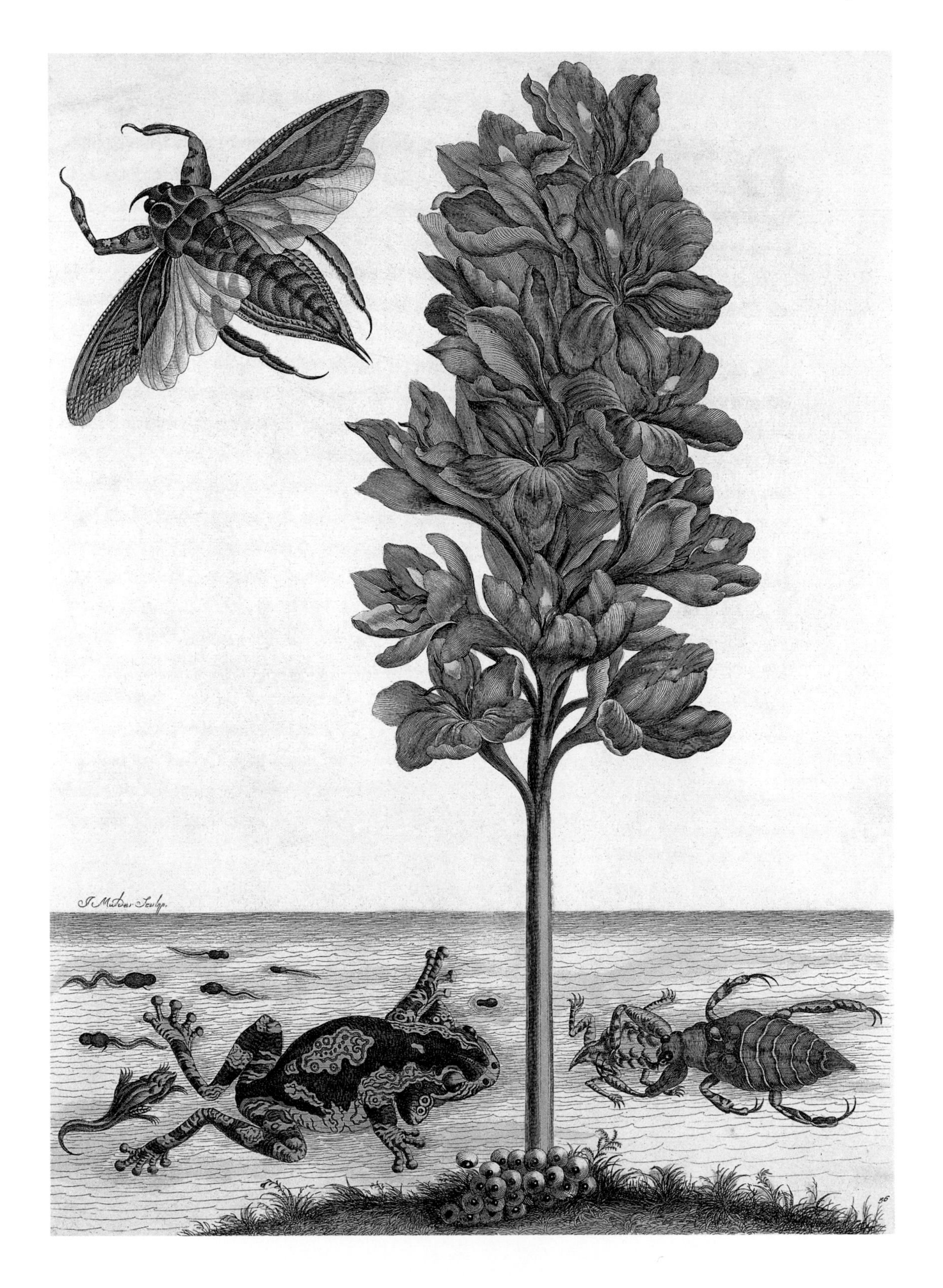
J. Mulder sculp.

Two hand-painted plates *(above & opposite)* from Merian's *Neues Blumenbuch*, or New Book of Flowers, published in 1680 and now extremely rare, hint at the artistic talents that would be fully formed by the time of her trip to Surinam. The plates are indicative of the popular aesthetic of the day, but her inclusion of insects hints at what was to come – a passion for entomology that would change the course of her life.

The leaves and fruit of **Citrus medica** *(above)* provide a resting place for two varieties of insect. Merian did not know the name of the 'beautiful black beetle' below but included it because of its rarity – it has since been named the harlequin beetle, *Acrocinus longimanus*. On the leaf above is the moth *Phobetron hipparchia* with its cocoon and larva. The caterpillar, chrysalis and adult moth of *Eumorpha labruscae (opposite)* sit on a branch of the vine *Vitis vinifera*. These plants, she wrote, with their 'blue, green and white Wyn-druiven [wine grapes]', grew so abundantly and quickly in Surinam that a crop could be harvested only six months after planting. Wine could then be made locally and no longer imported – a sufficient surplus could even be produced, she maintained, for export to Holland.

P Sluyter sculp.
34

J. Mulder Sculp
24

Two species of beetle *(opposite)* on a plant which is probably *Argemone mexicana* – the Mexican poppy, or prickly poppy. Above is a long-horned beetle, *Callipogon cinnamoneus*, with its fleshy larva in the centre of the plant. The larva and adult of *Taeniotes farinosus* are on the left-hand leaf and at the bottom. A large pineapple *(above)*, *Ananas comosus*, dominates this simple composition. Merian enthused about the fruit's taste and scent: '... the skin is as thick as a finger; if it is not peeled, enough sharp hairs remain on the flesh which prick one's tongue while eating and cause great pain. This fruit tastes as though one had mixed grapes, apricots, red currants, apples and pears and were able to taste all of them at once ...' The caterpillar and chrysalis on the right, and the butterflies on the left, belong to *Philaethria dido*.

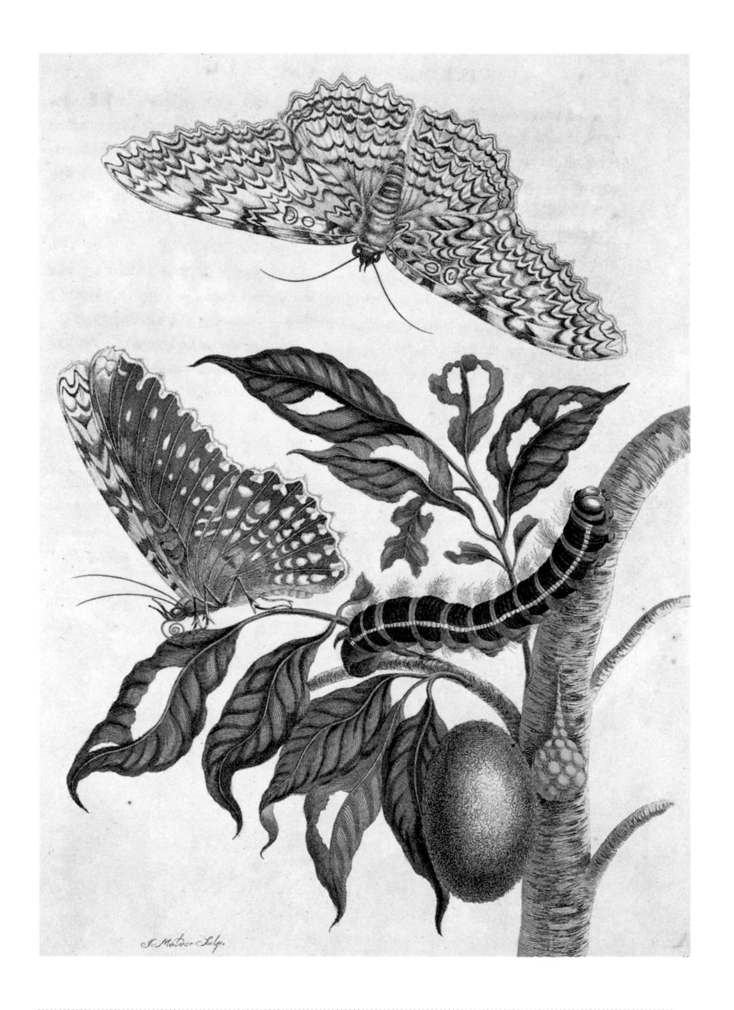

The caterpillar, cocoon and adults *(above)* of the *Thysania agrippina* moth. 'In the year 1700 during April I was in Surinam on the plantation belonging to the Misses Sommelsdyk called Providentia where I carried out various observations of insects …' Merian recorded. Here she found the caterpillars of *Thysania* on what she called a 'Gummi Guttae' tree – the species of gum tree shown here. An assemblage of unrelated insects *(opposite)* is dominated by a stag beetle in flight (*Macrodontia cervicornis*). At the bottom is an unidentified cocoon and an unnamed caterpillar which Merian described as resembling a 'clothesbrush', and mistakenly believed to metamorphose into the orchid bee, shown at the bottom. The grub and beetle of the palm weevil, *Rhynchophorus palmarum*, are seen in the middle. The grub was local gourmet fare: 'These worms are placed on charcoal to roast them and eaten as a very fine delicacy,' Merian wrote.

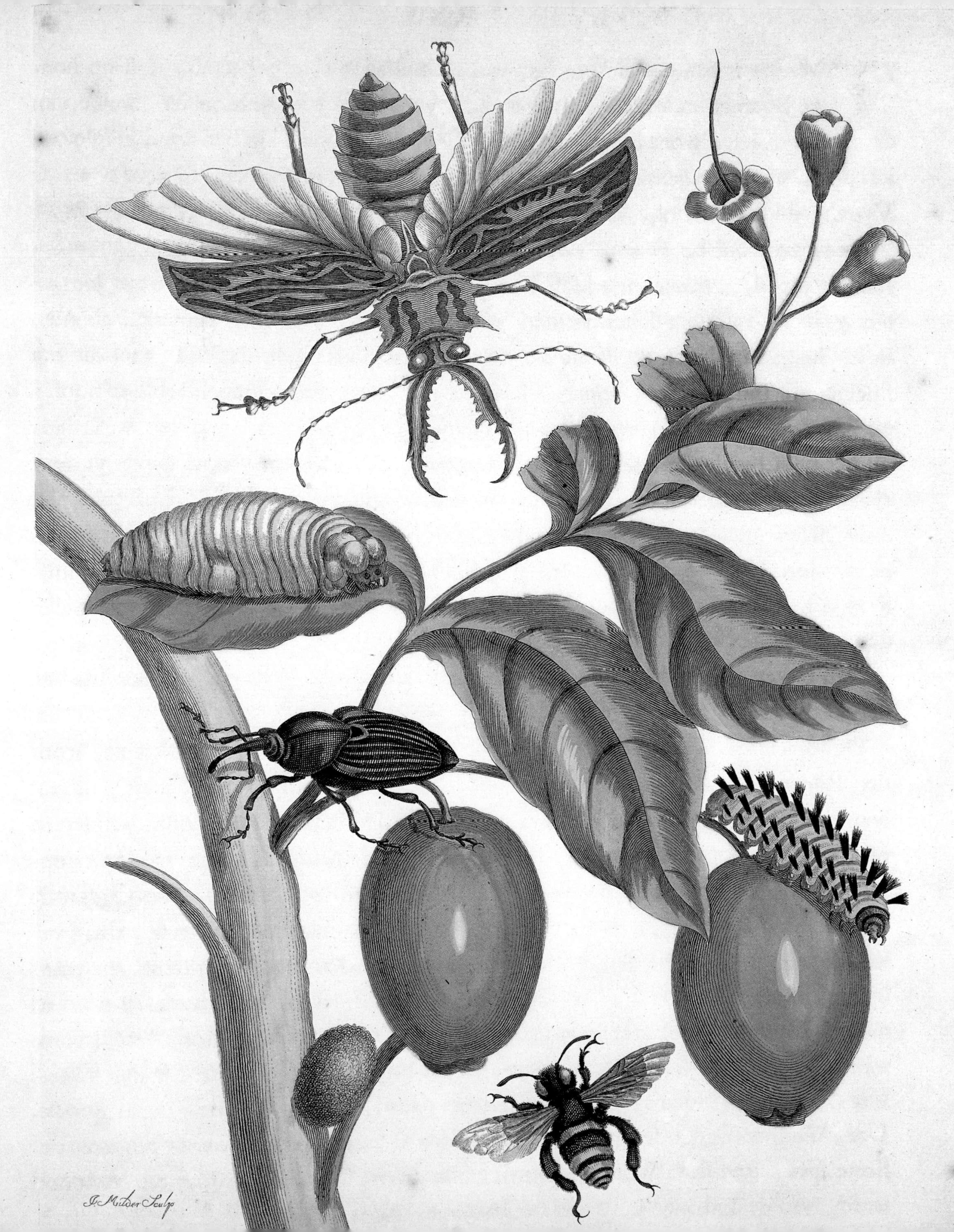
J. Mulder Sculp
48

1
1.
1.
2.
P. Sluyter Sculp.
46

***A green caterpillar and adult moth of* Erinnyis ello** *(opposite)* on a branch of Royal or Spanish jasmine, *Jasminum grandiflorum*. The caterpillar fed on the leaves of this and other plants. Merian observed that the snake, *Corallus enhydris*, at the base of the jasmine would curl itself up into tight coils and tuck its head out of sight inside them. Snakes, lizards and iguanas liked to hide beneath the jasmine bushes of Surinam which, Merian remarked, grew so profusely and thickly there that the air was heavy with their perfume. Despite her painstaking observations and recording of insects in their immediate habitat, some plates in her *Metamorphosis* were more decorative, such as this one *(above)* including insects, plants and shells against a landscape.

Each of the 60 plates in the 1705 *Metamorphosis* was produced from an engraving *(opposite)* made from Merian's watercolour. Merian even made some of the engravings herself. The reproduction *(below)* appeared in a 1981 facsimile edition of the book. In the engraving, the spider at top left is *Heteropoda venatoria*, which carries its eggs in a sac under its body – the newly hatched young can be seen here. Of the large black spiders, *Avicularia avicularia*, Merian wrote: 'When they fail to find ants they take small birds from their nests and suck all the blood from their bodies.' This last, fanciful detail is unlikely in reality, but Merian has illustrated it with an *Avicularia* feeding on a hummingbird.

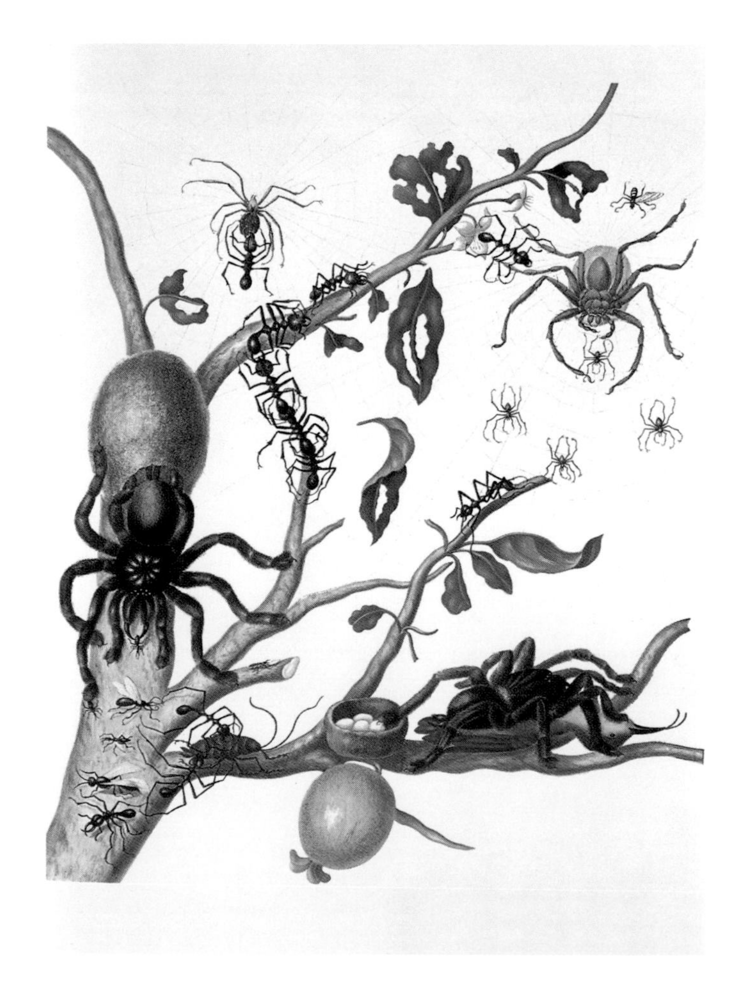

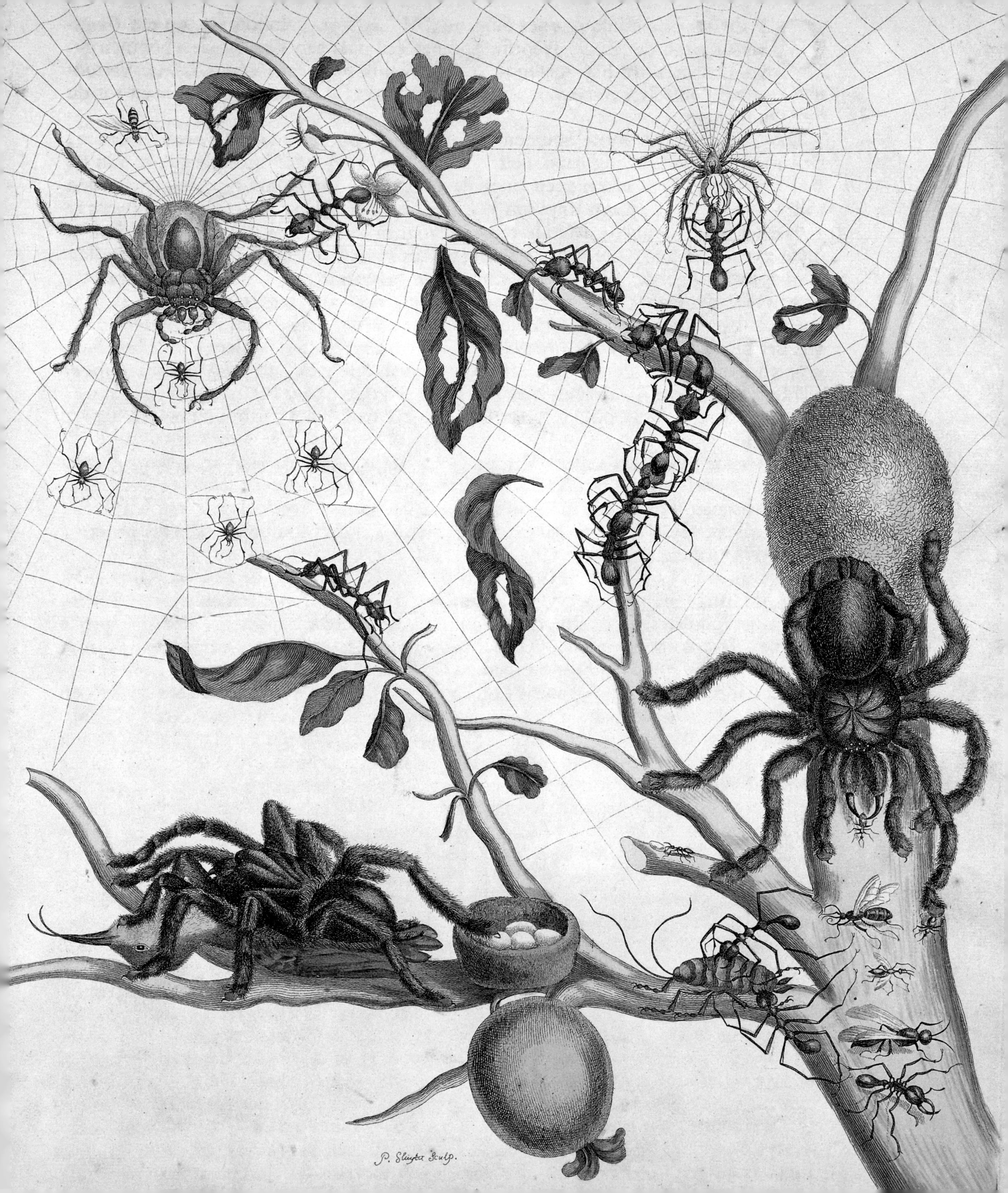
P. Sluyter Sculp.

***The butterfly* Morpho menelaus *(above)*.** Merian described it as '... looking like polished silver overlaid with the loveliest ultramarine, green and purple, and indescribably beautiful; its beauty cannot possibly be rendered with the paintbrush.' Nevertheless, she tried, using a fine, pointed brush to apply jewel-like colour to the finest vellum, known as *carta non nata*, the skin of unborn animals, which provided a much smoother surface than canvas, wood or handmade paper. A caterpillar and butterflies of *Heliconius ricini (opposite)* feed on *Ricinus communis*, the caster-oil plant. In Surinam its oil was used to treat wounds and was burned in lamps.

P Sluyter Sculp.

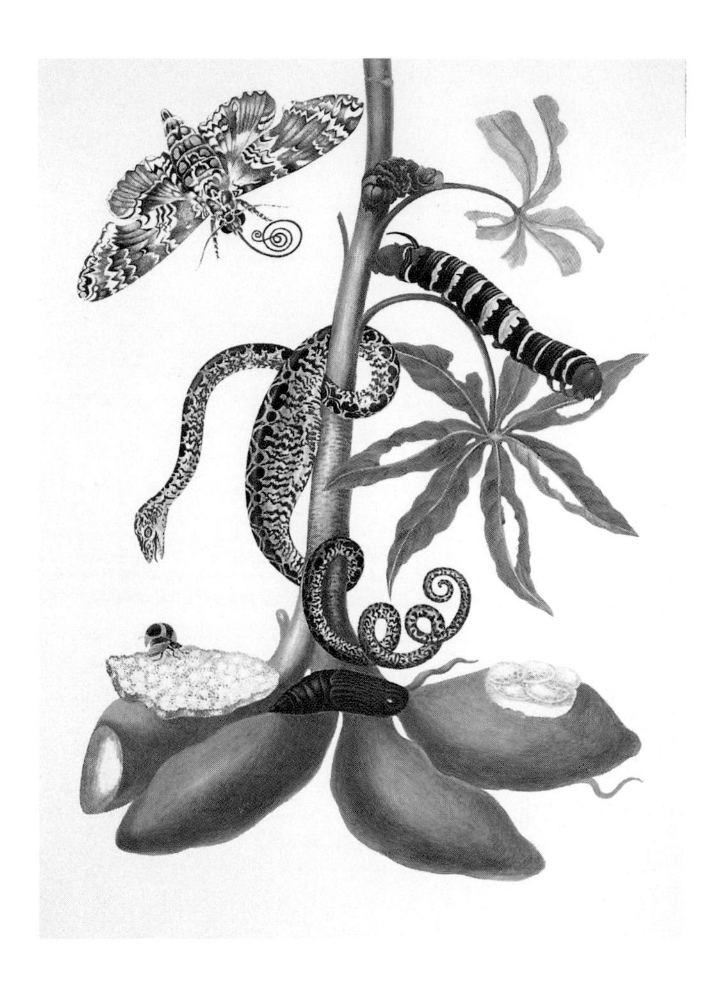

A cassava or manioc plant, *Manihot esculenta*, provides a visual framework for this composition *(above)*. 'The root is grated,' Merian wrote, 'then one presses out the juice, which is very poisonous; then the pressed out root is laid on an iron plate like those used by hatmakers in this country; under this plate one lays a small fire so that the remaining moisture can evaporate; it is then baked like a rusk and has the same taste as a Dutch rusk ...' The stripy caterpillar feeding on the plant was a great pest of the cassava fields, Merian noted. A tree boa, *Boa hortulana*, now known as *Corallus hortulanus*, with a potential length of 2 metres (7 feet), coils decoratively around the stem.

***A caterpillar and butterfly of* Caligo teucer** feed on a banana *(above left)*. The lizard, *Cnemidophorus lemniscatus*, was included for purely decorative reasons. It had built its nest in the floor of Merian's house, and laid four eggs there, three of which are shown on the stalk. She took the eggs with her on her return journey to Holland, during which they hatched but did not survive. The cotton leaf Jatropha, *Jatropha gossypifolia (above right)*, had medicinal properties that Merian learned about from the American Indians and African slaves: its root was used to treat snakebite, while its leaves were used as purgatives and in enemas. The large green caterpillar, the reddish chrysalis, the large moth – and even the empty skin from a previous shedding and the rounded faeces on the stem – all belong to the anteus moth, *Cocytius antaeus*, which feeds on Jatropha.

***After publishing* Metamorphosis,** Merian planned a second volume with illustrations of the reptiles and amphibians she studied in Surinam, but could not finance it in the end. Merian illustrates three stages in the life cycle of the mantid, *Stagmatoptera precaria (opposite),* on the branch of a tree. Below the tree, with its babies clinging to its back, is a creature which Merian referred to as a 'forest rat'. It is in fact an opossum, *Didelphys spp*. Framed by the arching stem of *Sesuvium portulacastrum (above),* a cress-like plant eaten locally, is the toad *Pipa pipa*, which was 'eaten by the natives who consider them to be a good dish'. Of its reproductive methods, she wrote: 'The female carries its offspring on its back and has its womb placed on its back; here it receives and develops the sperm; when the eggs have matured the young toads work their own way out of the membrane; they crawled out one after the other as though all came out of one egg. When I saw this, I threw the female and her young into brandy ...'

A composition *(left)* based on the studies Merian made of the reptiles and amphibians of Surinam. Here, she shows a Surinam caiman, probably *Paleosuchus palpebrosus*, biting what she called a 'viper', but what is now known as *Anilius scytale*, the South American false coral snake. Behind the caiman a young reptile emerges from an egg – this is not, as might be supposed, the young of the larger caiman, but another species, *Caiman crocodilis*. Merian was highly impressed with its growth rate. Coming from an egg 'as large as that of a goose', in a very short time it had become 'seven or eight times larger' than the egg out of which it hatched.

The reptile **Tupinambis nigropunctatus** *(overleaf),* commonly known as Tegu, was to be found in the forests of Surinam, Merian wrote. She had little else to say about it except that it was a species of lizard that ranked for size between the salamander and crocodile. Like other lizards, she said, it came from an egg, and she found from experience that it ate the eggs of birds.

Travels in North America 1753–1777

William Bartram

***A species of flowering treee** (opposite)* discovered by John and William Bartram on their expedition, which they named *Franklinia alatamaha* in honour of their friend and mentor, Benjamin Franklin.

THE YEAR 1753 WAS AN AUSPICIOUS ONE FOR THE DEVELOPMENT OF BOTANY because it saw the publication of Carl Linnaeus' *Species Plantarum*, the starting point for modern botanical nomenclature. Using the new system in which each plant species, no matter where it was from, would have a simple two-part Latin name, botanists would henceforth have the wherewithal to avoid the dreadful confusion that had bedevilled their subject in the past. Nowhere was the new system needed more than in North America where thousands of plant species, unknown to the rest of the world, remained to be discovered by science.

That same year, the 14-year-old William Bartram, who would subsequently make many of these new discoveries, accompanied his father John on his first major botanising expedition from the family home in Kingsessing, Pennsylvania, to the Catskill Mountains in New York. John Bartram (1699–1777) was an unpretentious Quaker, a self-educated farmer, but with a passionate interest in natural history, and particularly plants, which he passed on to several of his sons, including William. In 1729, John had started a botanical garden at his home which became a sort of specialist horticultural nursery supplying North American plants and seeds to customers in Europe, including professionals like Linnaeus, and eventually European plants to colonial buyers. Quite apart from the potential medicinal and commercial value of the New World plants, it had been fashionable in England since the mid-17th century for rich landowners to populate their estates with exotic species from around the world. North America had been a particularly popular source of such plants until the early 1700s when the Orient, and especially China, became the new fashionable focus of attention. But interest in North American natural history had been maintained and was stimulated by Mark Catesby's illustrated account of the *Natural History of Carolina, Florida and the Bahama Islands,* published between 1731 and 1747, the original illustrations for which are now in the Royal Library at Windsor Castle in England.

One of Catesby's patrons in England was the Quaker woollen draper and horticulturalist Peter Collinson who later assumed much the same role for John Bartram, including acting as a sort of agent, putting Bartram in touch with potential customers and supplying him with European plants for colonial plant enthusiasts. As a result, John's European business

2.
1.
3.
4.

Perched on a twig on the left of this drawing is the cardinal, *Cardinalis cardinalis cardinalis (opposite),* which Bartram described alternatively as the 'Red Sparaw' or 'Red bird of America' or 'Virginia nightingale'. The twig on which it sits is that of *Osmanthus americanus*, or devil-wood. Bartram saw this 'curious and sweet scented shrub' – which he called '*Olea Americana, foliis lanceolato-ellipticis, baccis atro-purpureis* Catesby', or the 'Purple berr'd bay' – growing by the Cape Fear River in North Carolina. The fish, swimming so incongruously through the scene, is unidentified.

developed rather well and, apart from plants, he sold about 20 sets of seeds – each comprising 100 different North American species, at five guineas a time – each year between 1735 and 1760.

John Bartram had travelled up the Hudson River and 'botanised' in the Catskills in 1742, more than a decade before he took young William to the same area, and over the years he made a number of other collecting trips. Nevertheless, most of his plants came from his home colony of Pennsylvania, or at least from the relatively narrow strip of the eastern seaboard occupied by the British colonies from New England to the recently founded Georgia. For in the first half of the century, the British foothold in North America was still rather tenuous, competing with strong French interests in Canada and the Mid West and with Spain in Florida. But the Treaty of Paris in 1763 at the end of the Seven Years War changed the situation significantly, with the ceding of Florida by the Spanish and the more or less complete departure of the French, except from Louisiana. In the new climate of apparent stability – soon to be shattered by the colonists themselves – the English King George III was prepared to show a little more interest in the colonies than had been customary in the past. Collinson thought that John Bartram's next goal, despite his relatively great age, should be to investigate the botany of Florida. With this in mind, and with the support of several other powerful allies, in 1765 Collinson managed to get John appointed King's Botanist in North America with an annual stipend of £50, although this was six times less than that of his short-lived predecessor in the post, William Young, an immigrant German nurseryman from Philadelphia.

John was by now in his late sixties and his eyesight was failing. Nevertheless, between July 1765 and April 1766 he undertook a major collecting trip in South Carolina, Georgia and northern Florida, which included travelling 640 kilometres (400 miles) along the St Johns River from Fort George to its source. William had shown considerable promise as an artist since his teens and accompanied his father, both to assist him and to illustrate the plants and animals that they encountered along the way. The expedition was a great success and the two Bartrams discovered a number of previously unknown plant species. The most spectacular was undoubtedly the beautiful flowering tree, *Franklinia alatamaha,* named for the family's influential friend – Benjamin Franklin. The Bartrams found it only in very restricted localities, and it was

soon to disappear even from these; for more than 200 years it has been known only from cultivated specimens, including those grown by John Bartram back in Pennsylvania. Somewhat ironically, the first box of plants resulting from the expedition was presented to the King by Collinson in February 1768, but following Collinson's death shortly thereafter the second box was presented by Benjamin Franklin in the months leading up to the American Revolution.

Already aged 26 at the beginning of the Florida trip, William had been something of a disappointment to his father up to this point. Collinson had encouraged his artistic efforts and had publicised William's early works among his English acquaintances. But John hoped that his son would have a rather more conventional career and initially sent him as an apprentice to a merchant in Philadelphia. This did not work out, nor did a stint working with his uncle at a trading post at Cape Fear in North Carolina. Moreover, William had turned down Benjamin Franklin's offer to teach him the printing trade, and Franklin's alternative recommendation, in light of the young man's undoubted and improving artistic skills, that he should become an engraver. Finally, during the Florida trip, William was so taken with the region that he persuaded his father to set him up as an indigo planter on the banks of the St Johns River, but the venture soon failed.

After Collinson's death, William came under the wing of yet another English Quaker, Dr John Fothergill (1712–1780), a physician and botanist and owner of the largest private botanical garden in England at Upton in Essex. From 1768 Fothergill employed William to provide him with plants, seeds and natural history illustrations, and also financed his amazing expedition to the south-eastern parts of what became the United States of America while he was away. His father was by now in his seventies and William made this trip without him. He left Philadelphia in April 1773 going first to Charleston and then on to Savannah by sea. For the next three and a half years, moving his base from time to time, he undertook a whole series of longer or shorter expeditions by boat, on horseback or on foot, and either alone or accompanied by one or more companions, often Native Americans to whom he became known as 'Puc-puggy', the Flower Hunter. In this way he explored the coastal region of South Carolina and the Cape Fear River, the Savannah River valley and across Georgia to Mobile, Baton Rouge and Pointe Coupé, 'a flourishing French settlement on the Spanish shore of

the Mississippi'. His extensive travels in Florida included the St Johns River which he had visited with his father years before, and large areas of the northern part of the state including the Seminole town of Cascowilla. William's own account of all of this, *Travels Through North & South Carolina, Georgia, East & West Florida*, was eventually published in Philadelphia in 1791 containing only a small number of rather uninteresting illustrations. But it is an amazing document that includes details of the landscape, climate, general vegetation, individual plants and animals encountered and the appearance, behaviour and lifestyle of the many Native Americans with whom he met and lived.

Travels is remarkable in other ways too. For one thing the chronology is extremely curious. At the end of his long absence, Bartram returned to Philadelphia in January 1777 only months before his father's death. But somehow he had become a little confused and had 'gained' a year during his travels. As a result, his book suggests that he actually came back in 1778. Moreover, his descriptions of some of his adventures, particularly encounters with Native Americans and animals, are not always strictly factual, for he seems to have been unwilling to allow the truth to get in the way of a good story. Understandably, therefore, the book received mixed and generally less than enthusiastic reviews, with doubts expressed even about some of Bartram's most faithful reporting, including his graphic accounts of the St Johns River alligators. This was naturally a great disappointment to him, though the work met with more approval from the wider readership. Over the 200 years since its first appearance, however, Bartram's book, and the travels on which it was based, have become increasingly revered, particularly in the USA – despite the idiosyncrasies, or perhaps because of them.

William Bartram's other rather modest wish – that he would receive recognition as being the discoverer of the many new plants that he had found – was not to be realised in his lifetime. During his association with Fothergill, he sent his patron some 209 plants, many of them new, and a total of 59 illustrations, though some of these were not completed until after Fothergill's death in 1780. Many of the drawings resulted from William's *Travels* and included excellent illustrations of birds, fishes, amphibians and reptiles and a variety of invertebrates as well as plants. While they are generally very accurate, several of them display a curious use of scale, so that the animals and plants depicted together seem grotesquely

'The Alegator of St Johns [opposite] Fig. 1 Represents the act of this terrable monster when they bellow in the Spring Season. They force the water out of their throat which falls from their mouth like a Cataract & a steam or vapour from their Nostrals like smoke,' was how Bartram, writing to Dr Fothergill, described the behaviour of the alligator in the foreground, in what is his most famous drawing. In his handwritten journal, *Travels in Florida,* he noted that sunrise was greeted by the alligator's 'dreaded voice'. The alligator received its formal name, *Alligator mississippiensis*, in 1801.

enlarged or diminished; moreover, some, like his smoke-belching alligators of the St Johns River, appear to be more fanciful than factual. Had they been widely publicised they would have given his detractors even more ammunition. As it was, the illustrations were not published in their entirety until 1968.

When Fothergill died, his collections, including Bartram's drawings and plant material, were acquired by Sir Joseph Banks and incorporated into his library and herbarium curated by Daniel Solander. But Solander was preoccupied with Pacific material – particularly from the voyage of the *Endeavour* – until his premature death in 1783. So the Bartram material remained largely unexamined until after the Banks collection eventually went to the British Museum in 1827, by which time William Bartram himself had been dead for four years.

Following his great travels, the second half of Bartram's life seems to have been comparatively quiet and without much public acclaim. In the first few years following the War of Independence, contact between the USA and Britain became rather more difficult than it had been in the past, though Bartram maintained his contacts with the American scientific community. He never married, but lived a bachelor's existence in his father's house, now owned by his elder and more business-like brother John who also managed the garden. Bartram helped out and showed the many visitors around the garden. He did a little illustration – to order – and was even for a time appointed Professor of Botany at the newly organised University of Pennsylvania, although there is no record of him ever lecturing there.

Strangely, after his great adventures in Florida and Georgia, Bartram did not travel much. This may partly have been due to his having badly broken his leg in a fall from a tree in 1786 while gathering seeds. But, in general, his early tendency to indolence seems to have come to the fore again in later life, so that in the last three decades of his life he appears to have done little other than gently vegetate. Having devoted so much of his life to satisfying the acquisitiveness of some of the most ardent collectors of his day, he accumulated remarkably little in the way of personal effects himself. His will suggests that at his death, he possessed only two chests of clothing, the feather bed and bolster on which he slept, two glasses and a tray, a tin letter box, some books and a purse containing some cash.

Fig 1

2

Eacles imperialis

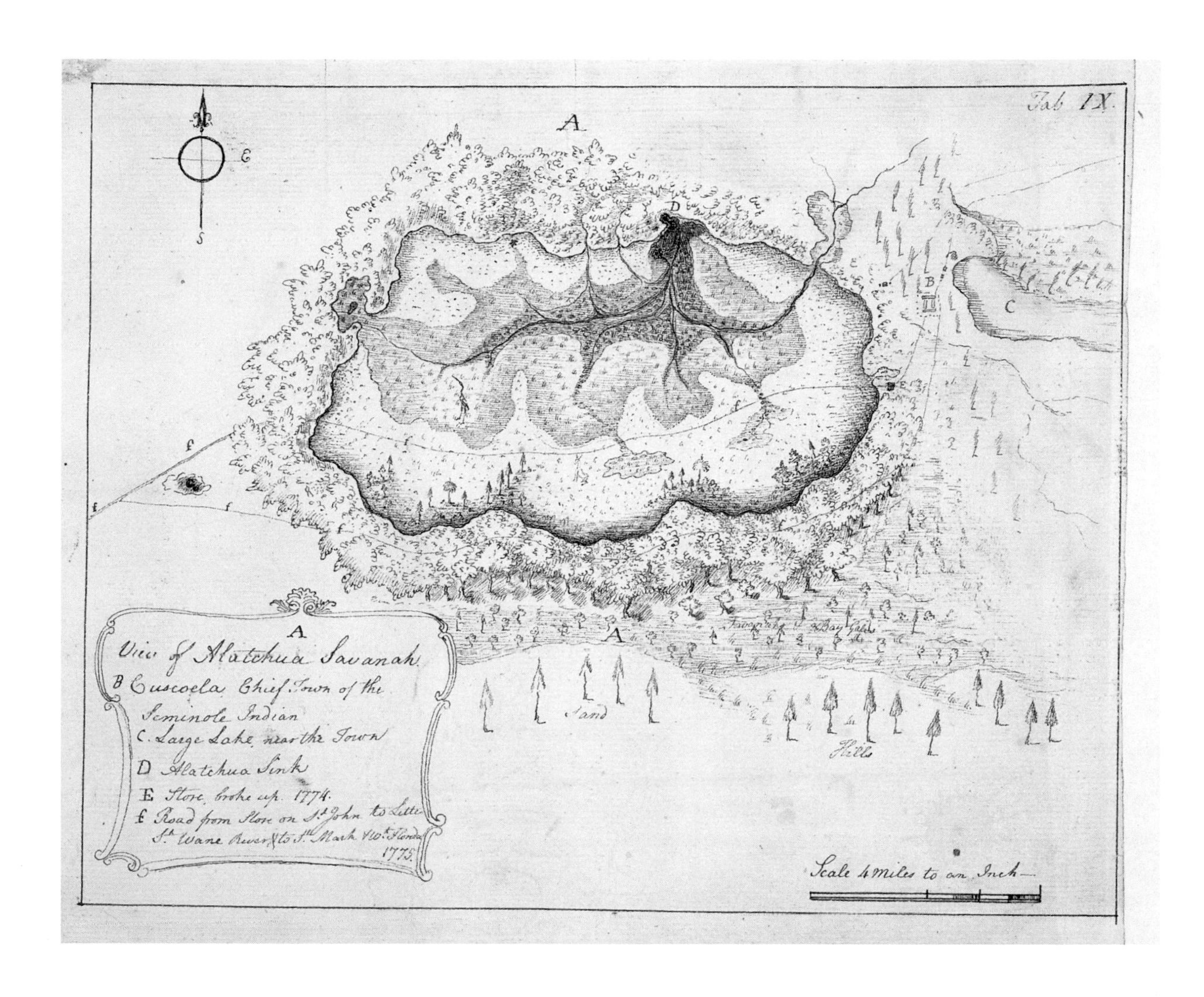

The savannah pink, *Sabatia bartramii (opposite)*, named after Bartram. He noted that it was 'an evergreen & flowers the year round. Is another lovely inhabitant of the Green Savanahs ... The flowers are produced in vast profusion very large, some I have seen when blown, 4 & 5 inches [10 to 12.5 centimetres] over of a deep rose color with a splendid golden star, in the midst rais'd on a bed of deep crimson. The number of Petals are various from 15 to 20 or 30 ...' One of the savannahs that Bartram explored, and drew, is now called Payne's Prairie *(above)*. It covers 34 square kilometres (20 square miles) south of Gainesville in Alachua County, Florida.

Bartram obtained the specimen for his drawing of *Momordica charantia*, the balsam pear or African cucumber *(above)*, from plants raised by James Alexander, a professional gardener and rival seed supplier to his father, John Bartram. His illustration of the American lotus (or water chinquapin), *Nelumbo lutea (opposite)*, is notable because it includes the first botanical sketch of *Dionaea muscipula*, the Venus fly-trap, tucked away to the left, under a lotus leaf. The bird striding in the foreground, identified as the great blue heron, *Ardea herodias*, (or possibly the sandhill crane) appears curiously out of scale when compared with the leaves and flowers of *N. lutea*.

Tab. I.

II
Fig 2

Like the drawing on the previous page, this one *(left)* also features the seedpod of *Nelumbo lutea*, the American lotus, together with *Triodopsis albolabris*, 'a large land Snale of a straw colour'd shell & pale inside. The snale when alive & without the shell is a very dark grey varied with black strokes & spots.' On the left is the waving flower of *Pterocaulon undulatum*, or blackroot – reputed to be very effective in treating dropsy and other diseases. Also shown on the left are the arrow-shaped leaves of Indian turnip or arrow arum, *Peltandra virginica* – whose root, Bartram informed Dr Fothergill, 'the Floridians roast or boil' – and water lettuce, *Pistia stratiotes*, which Bartram had seen growing with other plants 'all interwoven & matted together, forming vast marshes'.

***Writing in his* Travels,** Bartram describes how insects, lured inside the leaves of the insectivorous pitcher plants *Sarracenia purpurea* and *Sarracenia flava*, or trumpet leaf *(below)*, are trapped by a barricade of stiff, downward-pointing hairs which 'prevent the varieties of insects, which are caught, from returning, being invited down to sip the mellifluous exuvia [honeyed secretions], from the interior surface of the tube …' Unable to escape, the insects 'dissolve and mix with the fluid' that collects within the pitchers. In the same drawing, swallowing its prey, is *Cemophora coccinea*, the scarlet snake. In his rendering of *Canna flaccida*, or Indian shot *(opposite)*, Bartram includes a stone pipe bowl, thought to be the one presented to him by an old native American chief at Muklasa, by the Tallapoosa River in Alabama.

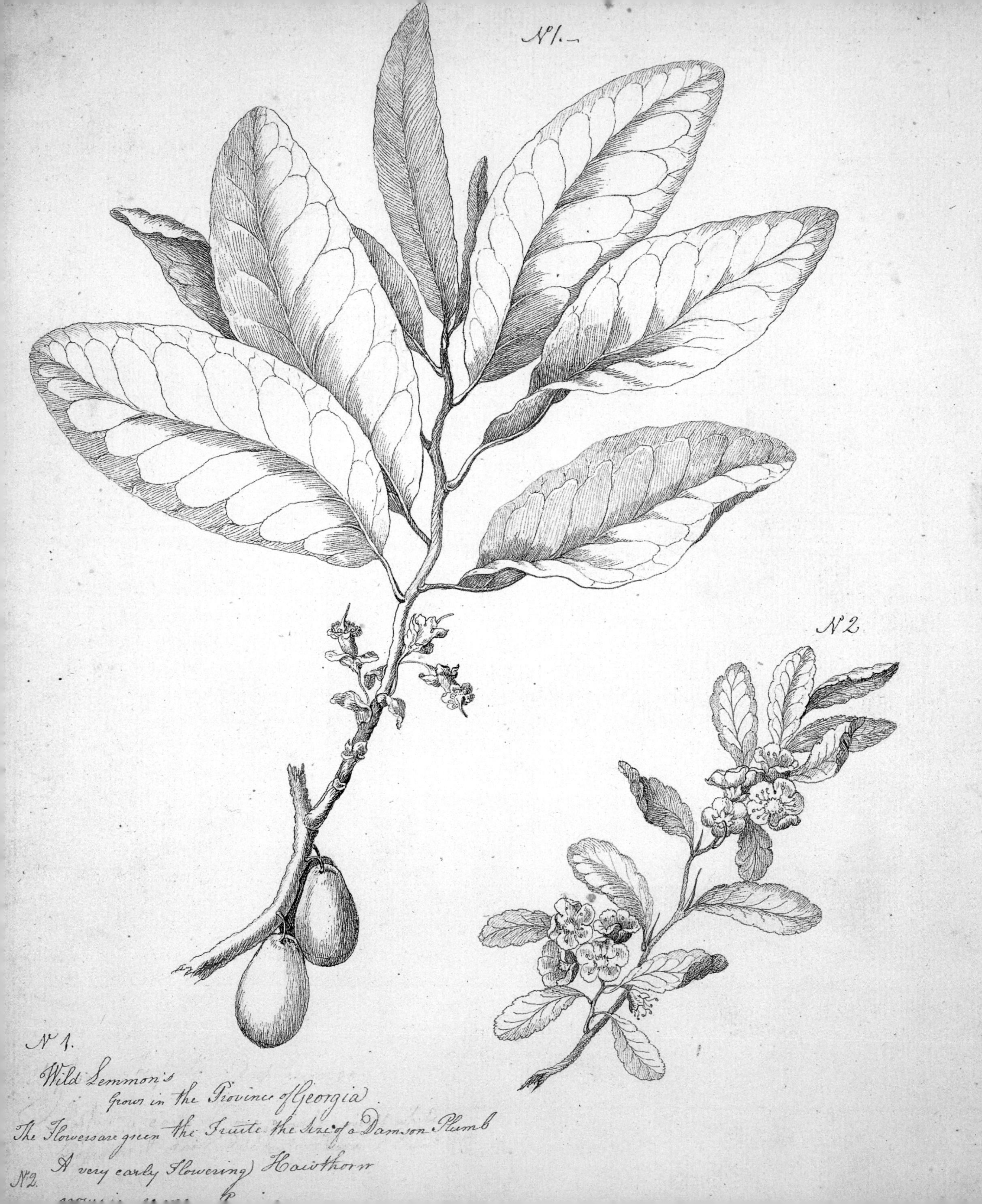

No 1.
No 2
No 1.
Wild Lemmon's
Grow in the Province of Georgia
The Flowers are green the Fruite the size of a Damson Plumb
No2. A very early flowering Hawthorn

As official Botanist to King George III, whose mother had founded the Royal Botanic Gardens at Kew, John Bartram was asked to send nuts from the American gum tree known as *Nyssa ogeche*, or ogeechee tupelo. A cutting of this plant was drawn by William Bartram *(opposite)*. In 1768, seed and plant importer Peter Collinson wrote to him from England, complaining, 'It's a little odd no tolerably good specimen is sent of it. I wish some nuts could be procured. There is no doubt of raising it.' The smaller cutting is from a 'very early Flowering Hawthorn' – *Crataegus sp.* – which Bartram also found growing in Georgia. The flowers *(above)* are probably from the insect devouring trumpet leaf plant *Sarracenia flava*.

Bartram's notes to this drawing *(above)* identify the bird shown as the 'Crested Red Bird of Florida or Virginia Nightingale'. This is *Cardinalis cardinalis floridanus*, not the same 'Virginia Nightingale' shown on page 123. Both birds belong to the same species, but are different subspecies. It seems to have been a constant inhabitant of the region for, in his 'Report to Dr. Fothergill', Bartram records that 'the Crested Red Bird [is] here all the Year'. He described the Florida sandhill crane, *Grus canadensis pratensis (opposite)* as the 'Wattoola Great Savanah Crane. Ash colour the quill feathers dusky or black'. The original specimen was sighted in Levy County, Florida, but was shot by Bartram's companions and eaten.

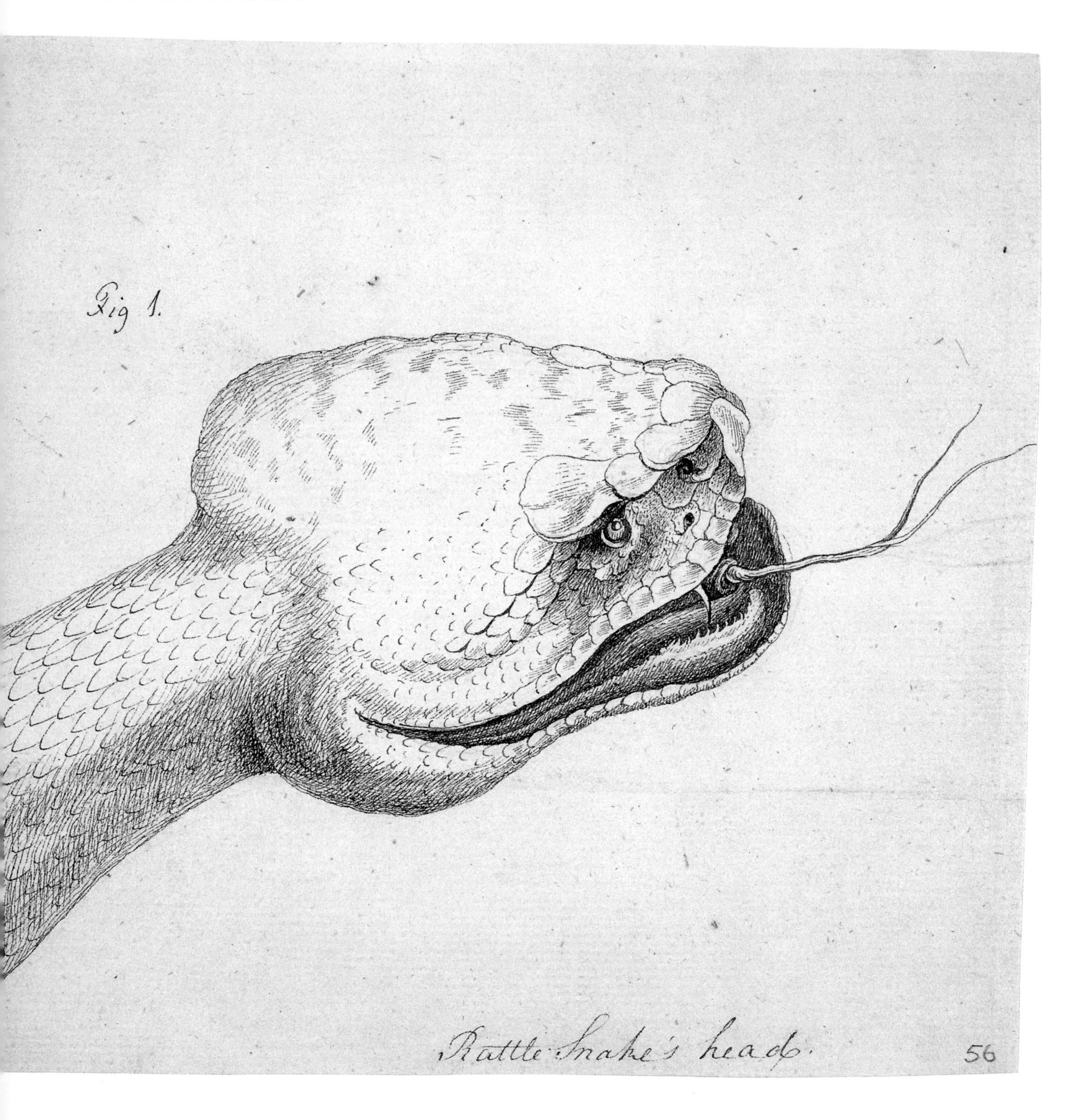
Fig 1.
Rattle Snake's head.
56

The 'great Rattle Snake' *(opposite)*, as Bartram called the eastern diamondback, the largest poisonous snake in North America. In *Travels*, he recounts killing 'a very large rattle snake' in a Seminole Indian camp: 'I took out my knife, severed his head from his body, then turning about, the Indians complimented me with every demonstration of satisfaction and approbation for my heroism, and friendship for them. I carried off the head of the serpent bleeding in my hand as a trophy of victory ...' Of the 'soft shell'd Tortoise of Georgia' *(below)* he wrote: '... the most singular in that tribe of Animals ... and when boil'd or baked may be eat with the flesh which is esteemed very wholesome & delicate food by the inhabitence of thet Province.'

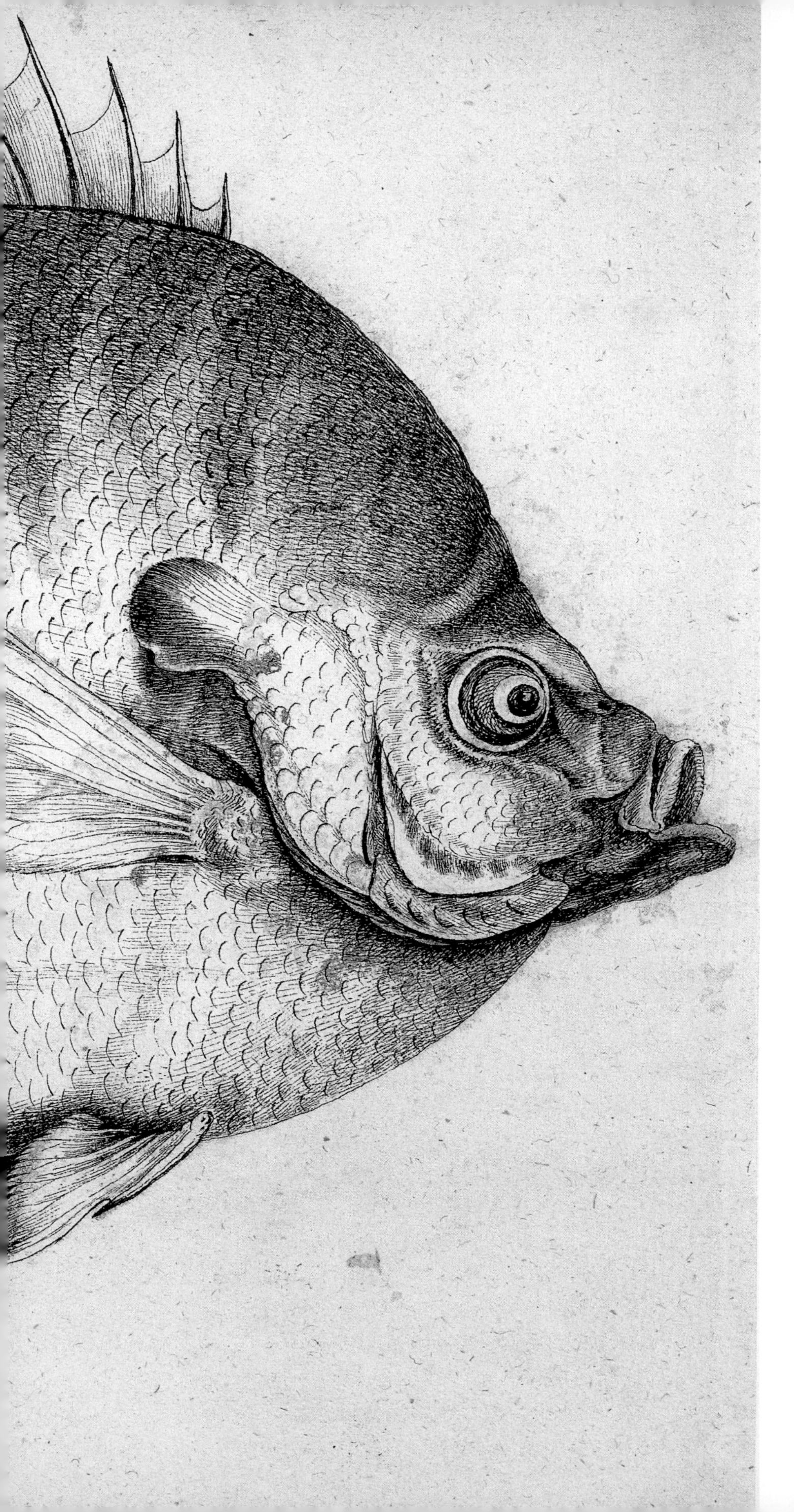

***Bartram saw this fish** (left),* his 'great black Bream', in East Florida. It is *Lepomis macrochirus purpurascens*, the copper-nosed bream. Bartram described it as being coloured deep purple, with large, black eyes flushed with red. 'This beautiful fish,' he reported, 'is plentiful in all Fresh water Rivers Sp[r]ings & holes in E[as]t Florida. His mouth appears remarkably small but by means of moveable Plates is capable of being extended wide enough to take young fish which is their prey together with Snales Periwinkles worms & aquatick reptiles. They are greatly esteemed as a delicious Fish.' Bartram referred to *Chaenobryttus coronarius,* or the warmouth *(above),* as the 'Great Yellow Bream calld Old Wife'. He had seen it in East Florida, noting that 'this is a very bold ravenous fish, like the leopard secretes himself in some hole or dark retreat & rushes out on a sudden snapping up the smaller fish passing by.'

Pacific Crossing 1768–1771

James Cook, Sir Joseph Banks & Sydney Parkinson

These first pictures of Aboriginals *(opposite)* through European eyes, made some 20–30 years after James Cook's voyage to Australia, belong to the Museum's collection of an unknown artist dubbed the Port Jackson Painter.

In the spring of 1768, James Cook, a 39-year-old Master in George III's navy, had every expectation that he was about to leave his pregnant wife and three young children to return to North America where he had served more or less continuously for the previous 10 years. He had built up an enviable reputation as a surveyor, usually undertaking the field work during the summer months and working on the charts back in England during the winter. It was this work to which he expected to return in 1768. But the Admiralty had other ideas; it was about to promote him to Lieutenant and give him command of the first of three great voyages which changed our knowledge of the Pacific.

The principal objective of this first voyage was to observe the transit of Venus, that is the passage of the planet across the face of the sun as viewed from the earth. Knowledge of the precise times taken for the transit as observed from different points on the earth's surface would enable astronomers to calculate, amongst other things, the distance between the earth and the sun. The last transit had been in 1761 – when, despite the efforts of 120 observers from nine nations, the results had been poor – and the next would not be until 1874. It was imperative to do better this time and all of the appropriate nations including, of course, the recently-at-war Britain and France, were anxious to participate. As part of the British effort, the Admiralty had agreed to send an expedition to make the observations from Tahiti, discovered only the year before by Captain Samuel Wallis in the *Dolphin*. But a voyage of such magnitude needed to satisfy more than one objective. Although the Pacific had been crossed many times in the previous two centuries, it was still largely unknown. At the very least, Cook was expected to discover new lands and take possession of any that promised benefit to the British crown. More particularly, he was to sail across the southern Pacific in search of the confidently expected, but so far undiscovered, *Terra Australis Incognita*, assumed to occupy much of the southern part of the globe as a sort of counterbalance to the mass of land in the northern hemisphere. Such a continent would surely abound with resources that would be an enormous asset to the first nation to locate and take possession of it.

The vessel chosen for the voyage, a 33-metre-long (106-foot-long) Whitby-built collier, suited Cook very well. Strongly built, extremely capacious, but with a very shallow draught, the ship was not fast, but was very manœuvrable

Ornamented after a Burial with a Club of great
the Shoulder

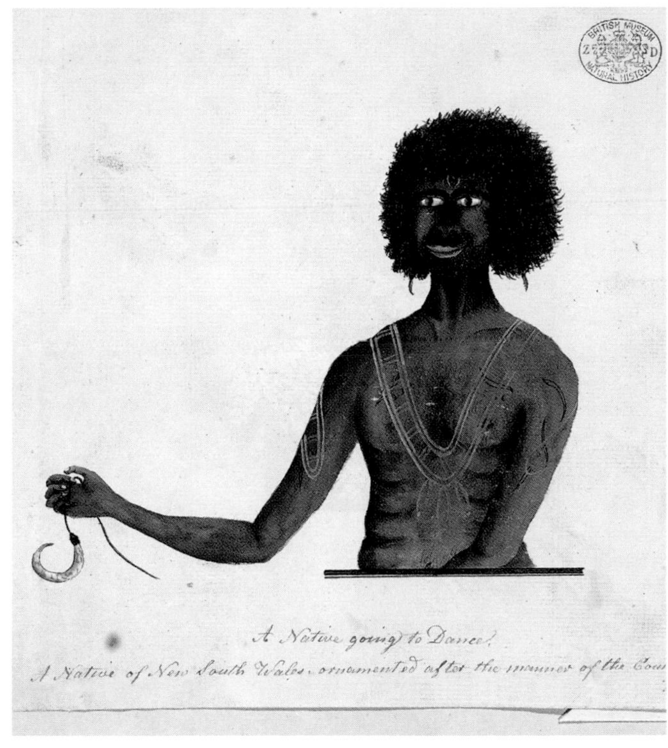
A Native going to Dance.
A Native of New South Wales ornamented after the manner of the

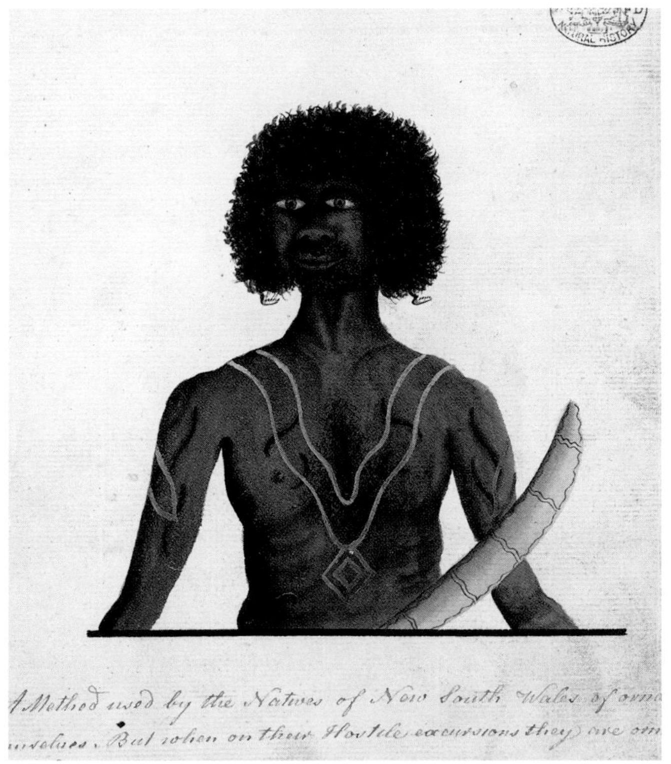
A Method used by the Natives of New South Wales of
But when on their Hostile excursions they are

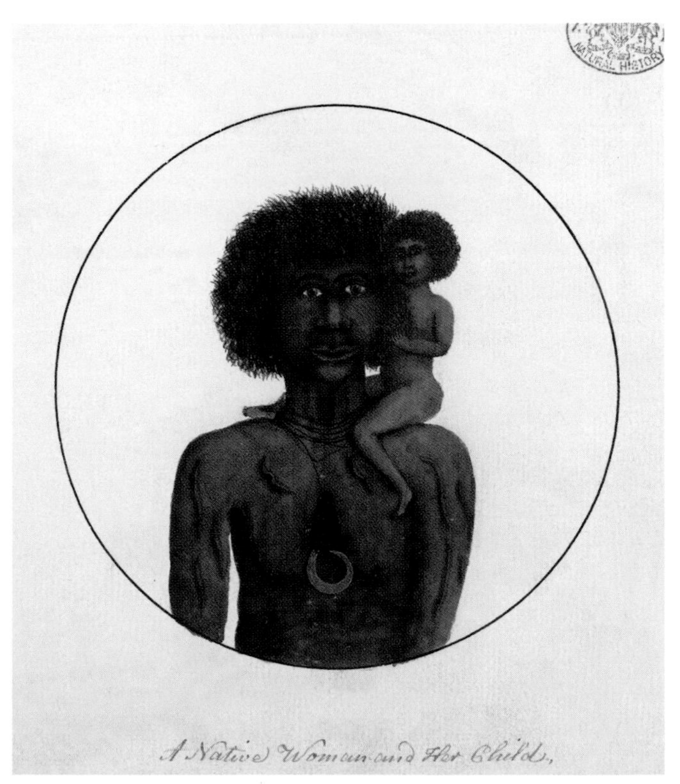
A Native Woman and Her Child.

The coastline of Sydney, stretching from Botany Bay to Port Jackson *(opposite top),* and south from Botany Bay *(opposite bottom).* The scenes are by Thomas Watling, a convicted forger who was transported to Sydney and painted much of the Port Jackson area for the colony's surgeon general, John White. Watling's notations 'Cape Banks' and 'Point Solander' (at the entrance to Botany Bay) refer to Joseph Banks and his botanist Daniel Carl Solander from the *Endeavour*. On the strength of his voyage to Sydney, Banks would later recommend its development as a colony.

and ideal for sailing in unknown and potentially dangerous waters. Less than four years old when the navy bought her, she was renamed *Endeavour* and refitted, her hull reinforced with a sheath of thin wooden planks heavily armed with large-headed nails as protection against the notorious shipworm of tropical waters.

Apart from accommodating the needs of her naval complement of 85, the *Endeavour* and her captain also had to accept an entourage of civilians, including an astronomical observer, Charles Green, to assist Cook with observations of the transit. Green had learned his trade as assistant to two previous Astronomers Royal and had accompanied the incumbent Nevil Maskelyne to Barbados in 1763 to test John Harrison's new chronometer. Harrison's invention was soon to revolutionise the determination of longitude by navigators, including Cook on his second Pacific voyage. But on this occasion, traditional methods of astronomical observation – more difficult and less reliable – were to be used, and Green was to prove invaluable to Cook in this respect.

Green's presence meant some reorganisation of the cabin spaces, but nothing like that required just a month prior to sailing when Cook was informed that he would also be accompanied by 'Mr Banks and his Suite', no less than nine persons in all. The 25-year-old Joseph Banks was handsome, rich, intelligent, well-connected and already a Fellow of the Royal Society. He saw the voyage as an exciting chance to further his study of natural history, particularly botany, and persuaded the Admiralty that he should accompany Cook at his own expense. He insisted on taking four servants, a secretary, two artists, a botanist, two dogs and a vast amount of baggage. Although finding room on the small and already crowded ship undoubtedly caused some upset, particularly among the more junior officers whose accommodation was directly affected, Banks and his retinue must take some credit for the overall success of the voyage. Certainly, nothing like the quantity of natural history and ethnographic material, particularly the extensive botanical collections accumulated by Banks and his botanist Daniel Carl Solander (1736–82), would have been collected in their absence. Banks' insistence on artistic coverage on this voyage also resulted in a superb visual record; and it set a precedent for taking dedicated artists on such voyages.

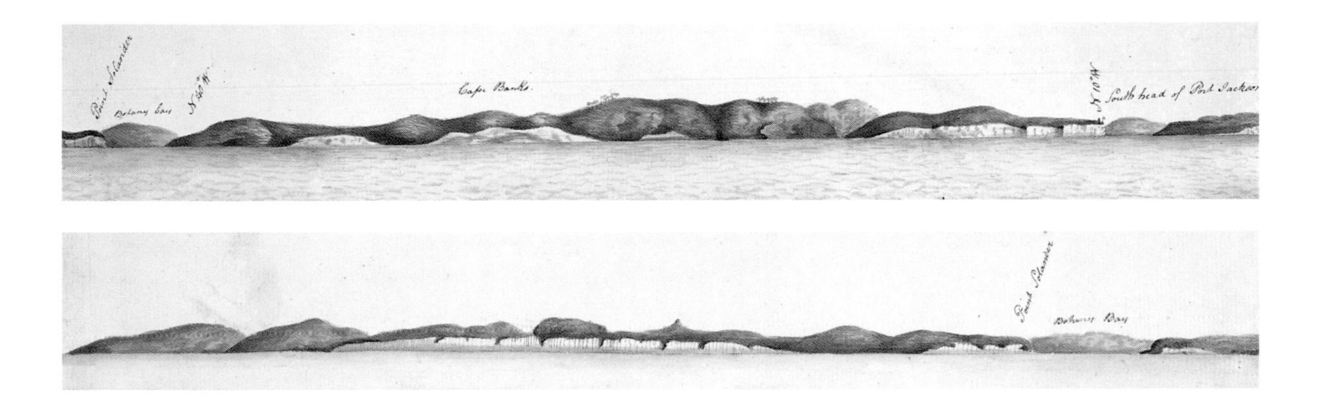

Banks' artists were Alexander Buchan, who was expected to illustrate people and landscapes, and the already well-known Sydney Parkinson (1745–71), who was to draw the plants and animals collected. Buchan, however, died a few days after the *Endeavour* reached Tahiti. As a result, Parkinson had to take on the enormous task of producing all the illustrations, with some help from Banks' Swedish secretary Herman Spöring, until both he and Spöring died within a couple of days of one another in January 1771 as the ship sailed across the Indian Ocean on the way home. The son of a Quaker brewer in Edinburgh, Parkinson had already been employed by Banks to illustrate natural history material collected during a visit to Newfoundland and Labrador in 1766 and to copy some of Pieter de Bevere's illustrations of Ceylonese animals. But he had never had to cope with the problems of painting on a heaving ship, nor of working in the discomfort of tropical climates.

The voyage, however, began smoothly. Cook left England in August 1768, called briefly at Funchal in Madeira before arriving at Rio de Janeiro on 13 November for a three-week stay. From Rio, Cook sailed direct to the Strait of Le Maire to round South America, and landed shore parties in Tierra del Fuego before finally passing Cape Horn and entering the Pacific in late January 1769. The *Endeavour* then sailed more or less continuously to the north-west, arriving at Matavai Bay, Tahiti, on 13 April, seven weeks before the transit was due. The observatory was set up, the transit successfully observed, and the ship made ready to sail by mid-July. Parkinson was showing remarkable resilience to the rigours of life in the tropics. Among his many problems were the Tahitian flies which, according to Banks, '... eat the painters' colours off the paper as fast as they can be laid on, and if a fish is to be drawn there is more trouble in keeping them off [it] than in the drawing itself. Many expedients have been thought of, none succeed better than a mosquito net which covers chair, painter and drawings, but even that is not sufficient, a fly trap was necessary to set within this to attract the vermin from eating the colours.'

From Tahiti, Cook sailed south to beyond 40°S, fulfilling instructions in his search for the southern continent. Having found no sign of a major land mass he turned first north-west then south-west and finally westwards to hit the

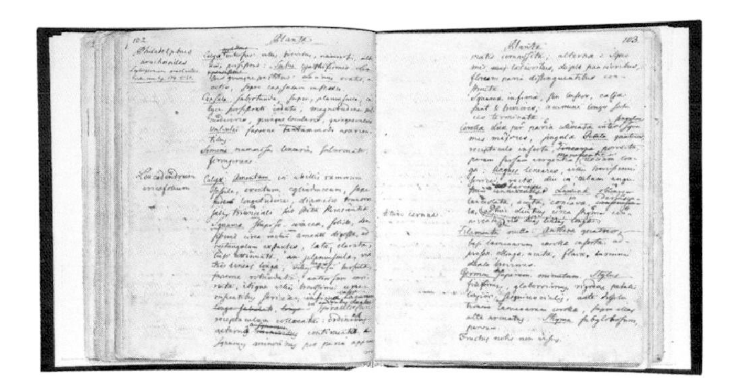

eastern coast of Abel Tasman's 'New Zeland', the western seaboard of which the Dutchman had come across in 1642. Cook sailed anti-clockwise round the North Island, then through the Cook Strait and clockwise around the South Island. Although several short land excursions were made, the remarkably accurate chart Cook produced was mainly the result of a 'running survey', that is, by sailing along the coast and taking thousands of careful bearings of land features and hundreds of astronomical observations.

Cook finally left New Zealand on 31 March 1770. Having achieved both principal objectives of the voyage, he could now have made his way home either back via Cape Horn or round the Cape of Good Hope. Since it was too late in the season to undertake the easterly route, Cook settled on the latter alternative, but decided that on the way he would chart the east coast of Australia northwards from where Tasman had left it 130 years previously. So began one of the most remarkable pieces of navigation and charting even by Cook's own high standards. At the same time it produced a wonderful collection of natural history material and some of Parkinson's best illustrations.

Arriving at the south-eastern tip of the continent on 19 April, the *Endeavour* sailed more than 2,000 miles along the whole of the east coast, including the dangerous shoals inside the vast Great Barrier Reef, at one point spending an agonising 36 hours aground on the reef and all but being destroyed. Apart from the excellent charts produced over five months of extremely difficult and hazardous navigation, several land excursions were made, by far the most important being south of present-day Sydney, in Botany Bay, so-named because of the wealth of new plants collected there.

Parkinson had already been busy since the beginning of the voyage, drawing marine animals and sea birds shot and collected from the ship during the ocean passages, and land plants and animals during the landfalls. But off the east coast of Australia he was overwhelmed. Banks and Solander were bringing in new material almost every day and Parkinson had to work frenetically to try to keep up, often working long into the night in cramped conditions by the flickering light of candles and oil lamps. He produced more than 400 sketches of plants while the ship was in Australian waters, though very few were actually completed during the voyage. Instead, his technique was to make detailed sketches, sometimes

Pages from the journal *(opposite)* of botanist on the *Endeavour*, Daniel Carl Solander, which he wrote on board. He kept a meticulous record of the plants collected on the voyage, so impressing Joseph Banks that Banks employed him for the rest of his working life.

coloured, of important parts of each plant, presumably for completion later in conjunction with the dried specimen. In contrast, many more of his animal paintings were completed, though these were mostly of fishes and birds from the earlier part of the voyage and not his most celebrated Australian sketches of kangaroos, made in June 1770 near what is now Cooktown in Queensland. In fact, Parkinson illustrated very few mammals; only the kangaroos and a quoll were from Australia. This is perhaps not altogether surprising. Plants could be kept relatively fresh for some time wrapped in wet cloths, and small invertebrates and even fishes would not 'go off' too rapidly. But a warm-blooded mammal was another matter – especially given that Parkinson drew in the ship's Great Cabin, where the officers and civilians had their meals.

By the time the ship reached Cape York at the north-eastern tip of the Gulf of Carpentaria her hull and rigging were in an appalling state and Cook decided that a call for repairs at the Dutch base of Batavia (present-day Jakarta) was essential before they attempted the journey home. The *Endeavour* arrived at Batavia on 11 October and left on 26 December 1770, battling against contrary winds for a further three weeks before they could finally leave the land astern. Although the ship was now seaworthy, the stay in Batavia had been disastrous to the health of the crew. When they left, Cook wrote that 'Batavia I firmly believe is the Death of more Europeans than any other place upon the globe of the same extent. We came here with as healthy a ship's company as need go to sea and after a stay of not quite three months left in the condition of a Hospital Ship, besides the loss of seven men.' Things were to get much worse, for there were to be a further 23 deaths, including those of Spöring, Parkinson and the astronomer Green, mainly from malaria and dysentery, before they reached Cape Town on 15 March 1771.

The month-long stay in Cape Town was recuperative, all but three of those that arrived sick recovered, and new crew members were recruited. Finally sailing on 15 April, after a fairly uneventful passage, the *Endeavour* anchored in Plymouth on 12 July 1771. The crew members naturally received an enthusiastic welcome home, not least because in their absence they had several times been given up for lost. Cook was praised by both the Admiralty and the scientific establishment and was promoted to Commander in August. But Banks was lionised. Along with Solander, he became the toast of

Self-portrait by Sydney Parkinson *(opposite)* from the art collection of The Natural History Museum. Despite his lack of formal training as an artist, Parkinson came to the attention of Banks when he first began to exhibit his paintings in London.

London society and his fame spread far and wide. Carl Linnaeus was so impressed by the natural history collections made during the voyage, including more than 1,000 plant species previously unknown in Europe, that he thought New South Wales should be named Banksia in his honour. Instead the name was given to a genus of plants.

In view of the obvious importance of the collections, and with Linnaeus' recently published *Systema Naturae* and *Species Plantarum* as superb guides, it is surprising that volumes on the new species obtained during the expedition were not published swiftly. Banks certainly intended to publish comprehensive accounts of the botanical material, and probably also the zoological collections – not only from this expedition but also those from Cook's second and third voyages, which he also acquired. But it never happened, mainly because Banks became more and more involved in other matters, particularly in his role as President of the Royal Society, a post he held for 41 years. Fortunately, the botanical specimens from all three voyages and most of the natural history illustrations were kept together and eventually came to the British Museum after Banks' death. Banks, however, was less interested in the zoological material which, as a consequence, was given away or sold to a large number of private or institutional collectors both in Britain and abroad, much of it being destroyed or lost in the process.

Nevertheless, Banks tried to do justice to the almost 1,000 finished and unfinished botanical paintings and sketches Parkinson had left, together with several hundred illustrations of zoological and other subjects. Despite spending more than £7,000 and employing 18 engravers to produce 753 plates from Parkinson's originals, Banks failed in his lifetime to have them published. Aside from 318 lithographs of Australian plants published at the beginning of the 20th century, and a selection of his aesthetically most interesting botanical illustrations published in 1973, it was not until 1980 that Banks' *Florilegium* was printed and justice to Parkinson's efforts on the *Endeavour* voyage was finally done. Far fewer of the zoological illustrations have been published, though many of them were used by later naturalists as a basis for the description of new species. Nevertheless, the scientific and artistic significance of the most important achievements in the tragically short life of Sydney Parkinson are now fully recognised, albeit more than 200 years after the artist's death.

***Daniel Carl Solander** (above right)* was engaged by Banks in 1768 at a salary of £400 a year to join him on the *Endeavour* as botanist. Part of his job was preserving the specimens collected on the voyage, including this one, *Marsilea polycarpa (opposite),* gathered by Banks. Solander's notes on the plants are contained in his *Plantae Novae Hollandiae (above left).*

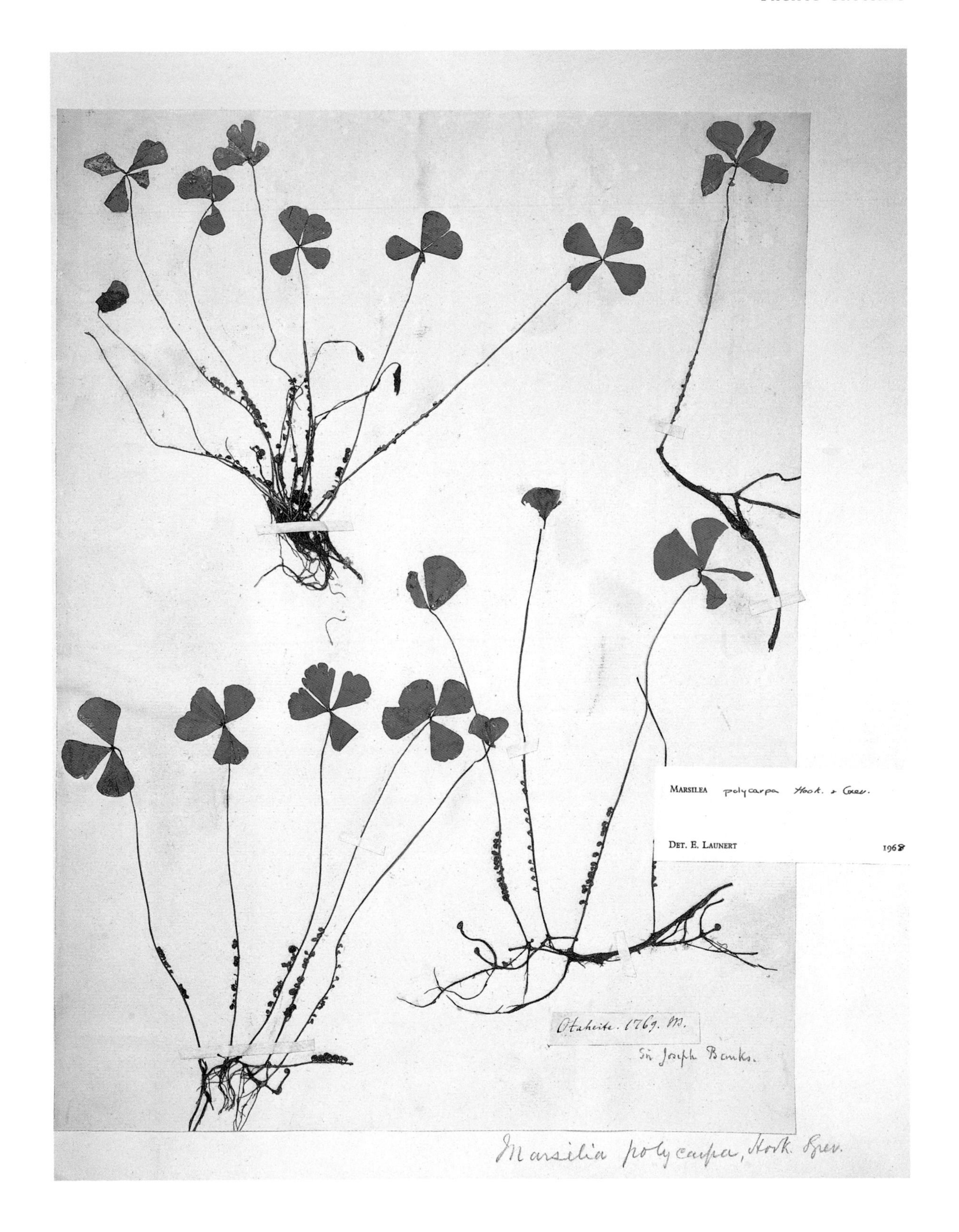
MARSILEA polycarpa Hook. & Grev.
DET. E. LAUNERT 1968
Otaheite. 1769. M.
Sir Joseph Banks.
Marsilia polycarpa, Hook. Grev.

Xylomelum pyriforme Smith

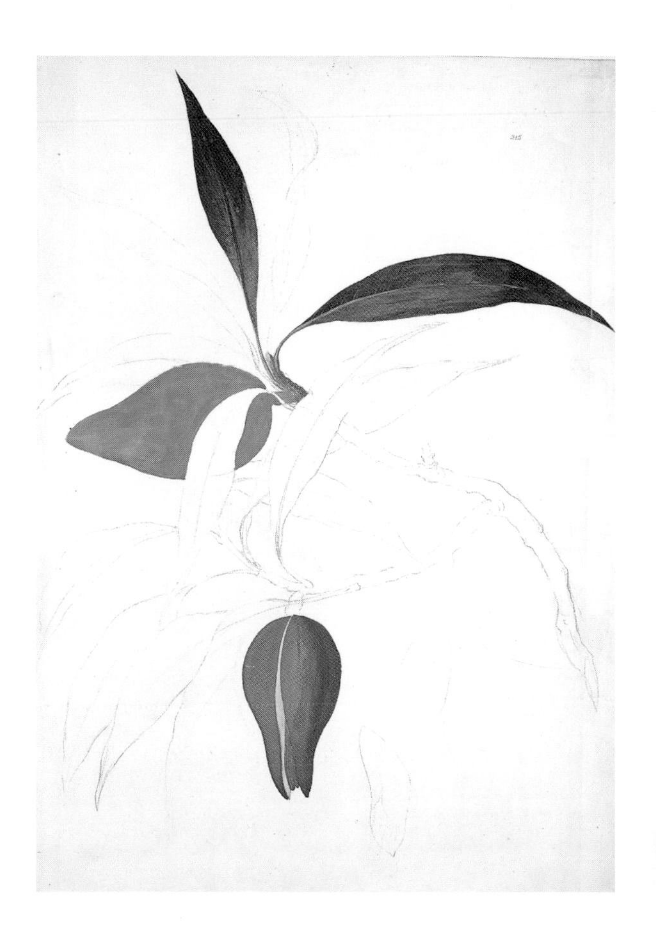

On his return to England, Banks planned to publish a book of engravings of the plants collected and painted by Parkinson. This involved several steps for each image, as these versions of *Xylomelum pyriforme* demonstrate. The original, partially coloured drawing made by Parkinson on the *Endeavour* voyage *(above left)* was copied and completed in London by one of several artists working on the project – John Frederick Miller – using watercolours *(opposite)*. An engraving plate was then made from this finished painting, and a single-colour proof *(above right)* made to check the engraving before the full-colour version was printed. The finished watercolour and engraver's proof are also shown for *Deplanchea tetraphylla (overleaf)*.

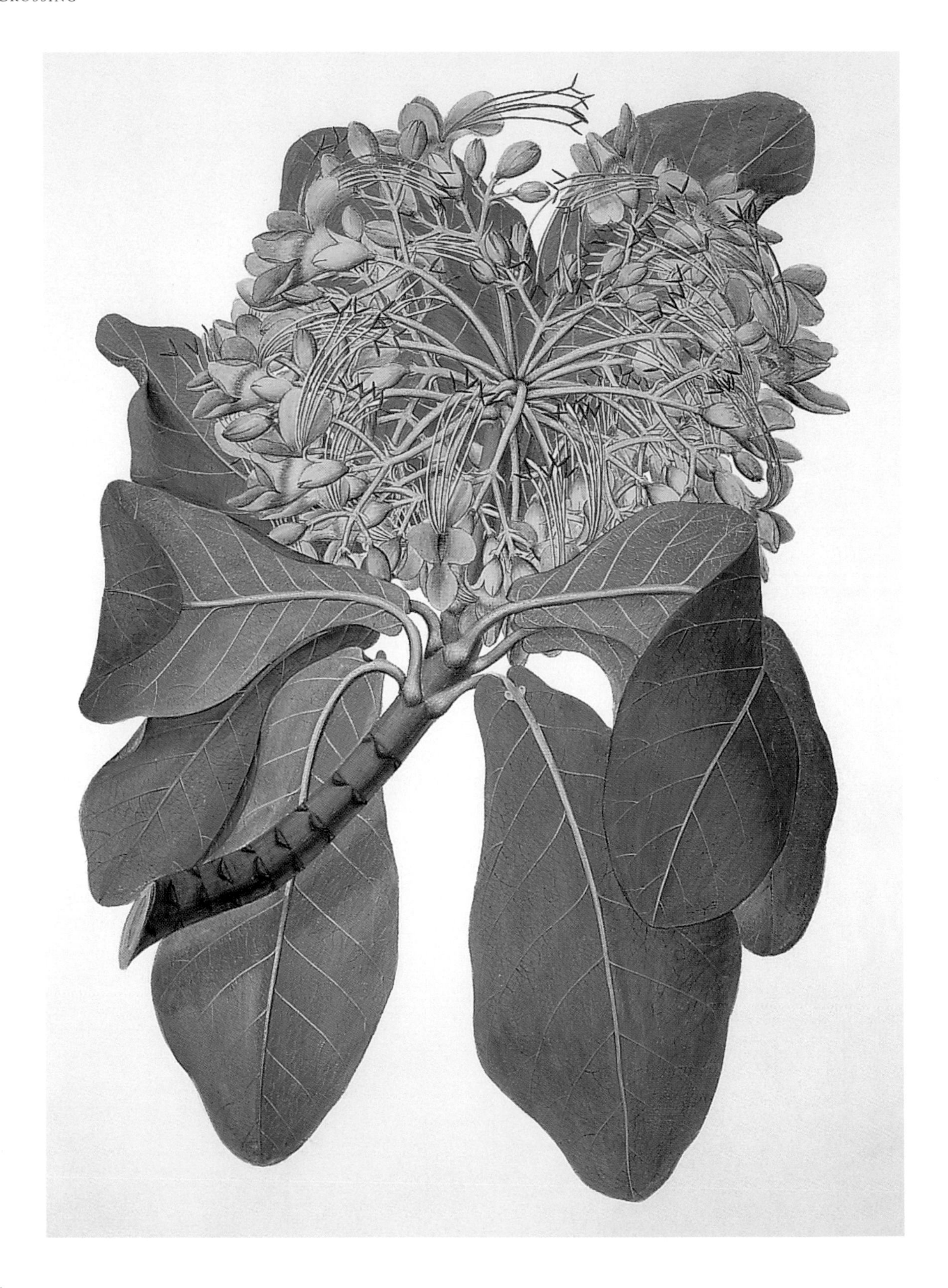

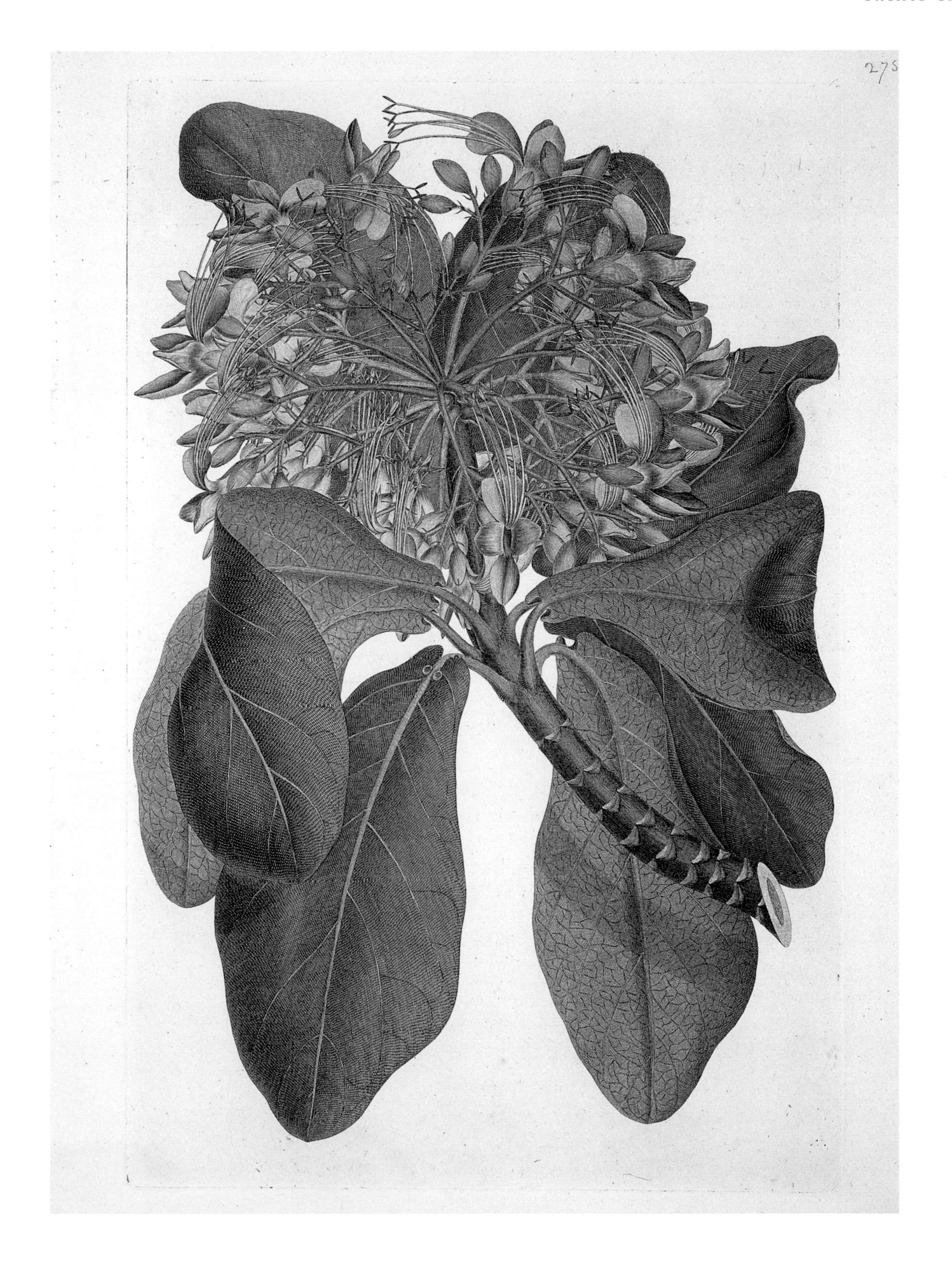

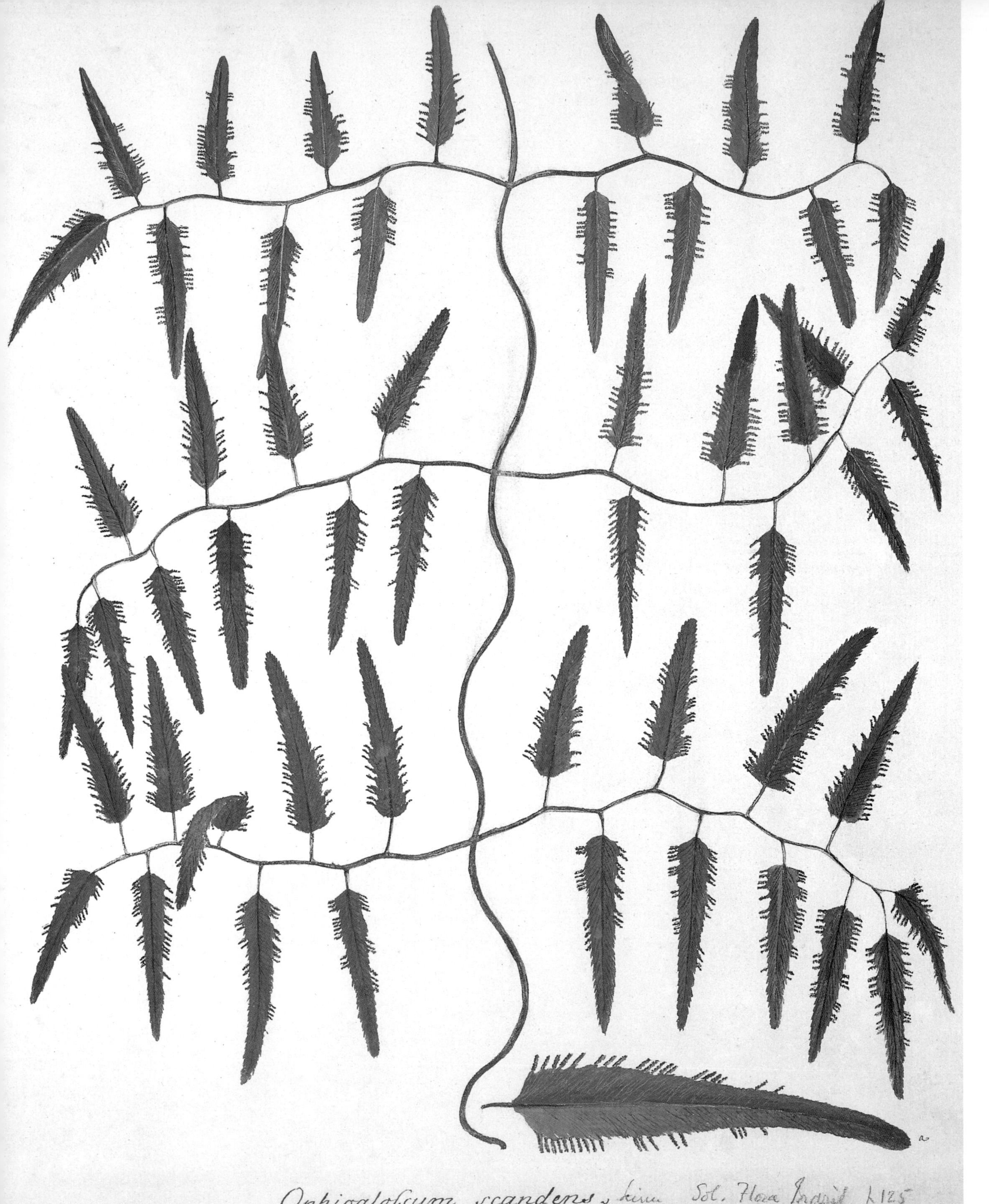
a
Ophioglossum scandens. Linn Sol. Flora Indic. p 125
Lygodium volubile Sw

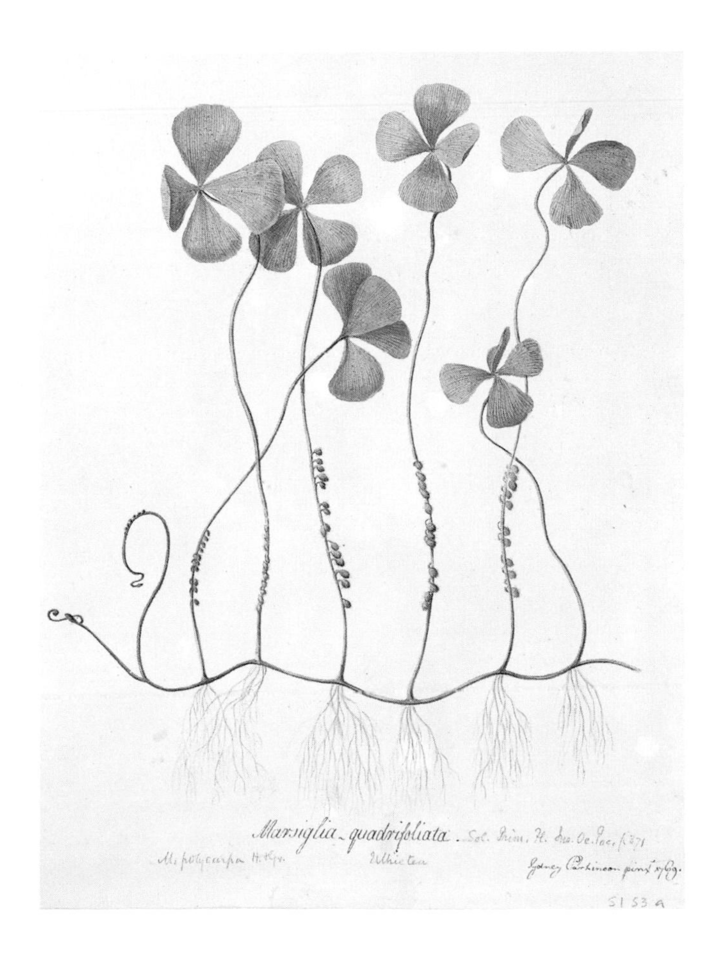

A climbing fern from Brazil *(opposite)*, named *Ophioglossum scandens* by Solander but known nowadays as *Lygodium volubile*. The plant was used traditionally in basketry and to make yarn, mats and fish traps. Although unsigned by Parkinson, the illustration is probably by him. Some 690 of the illustrations attributed to him are unsigned, but descriptive notes written in Parkinson's hand suggest that the bulk of these original works (not the later copies produced by artists such as Nodder and Miller) were by him. A signed Parkinson watercolour *(above)* is inscribed 'Sydney Parkinson pinxt 1769'. It shows *Marsilea polycarpa*, a type of water plant found in Tahiti, where the insects ate the paint from the canvas so voraciously that Parkinson and his easel had to be shrouded in mosquito netting.

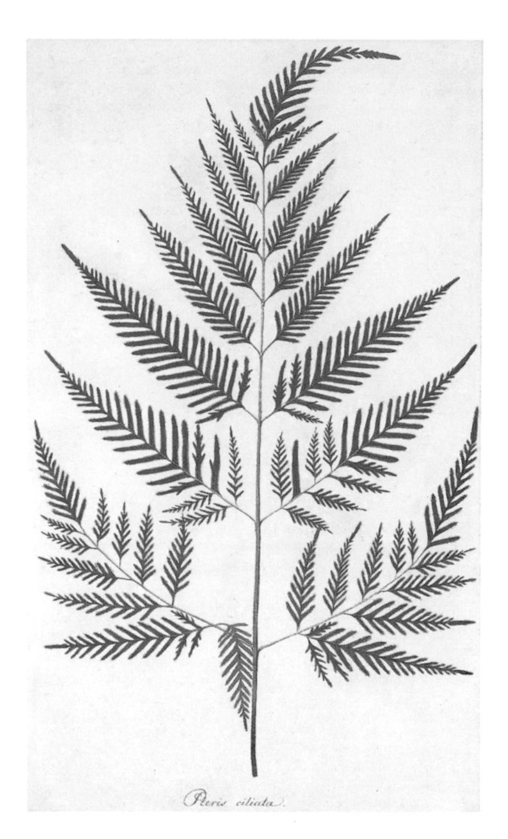

Several artists contributed to Banks' planned publication (eventually called *Banks Florilegium*). Engravings were made from Parkinson's originals such as the fern *Pteridium aquilinum (above left)* and sometimes from the works of unknown artists *(above right)*. Many of the watercolours were executed by the Miller brothers, John Frederick and his brother James. The watercolour version of Parkinson's drawing of *Banksia ericifolia (opposite)* – the genus *Banksia* was named after Banks – is inscribed 'John Frederick Miller pinxt 1773'. The species was found at Botany Bay, so-called because of the rich variety of plant to be found there.

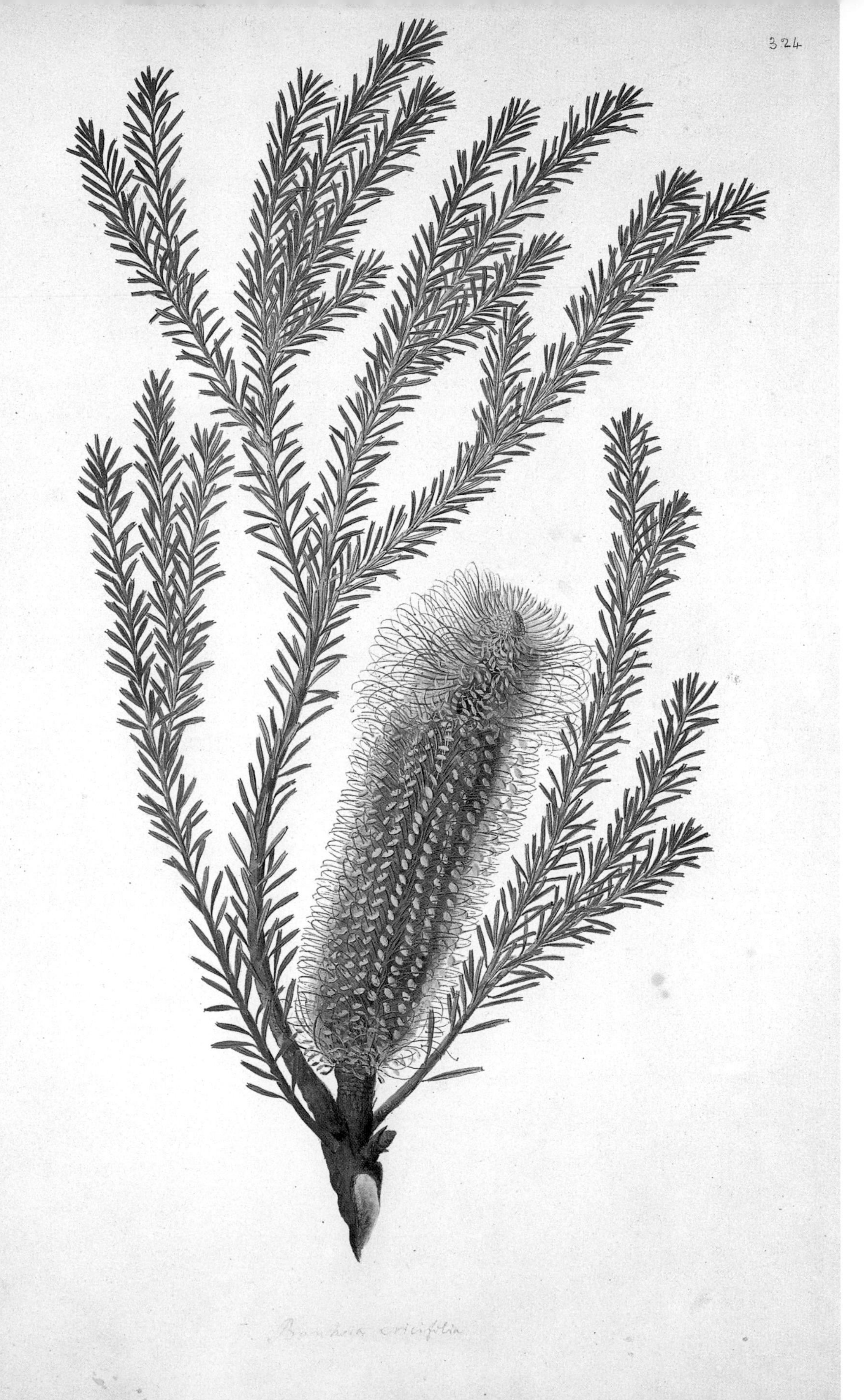
324
Banksia ericifolia
John Frederick Miller pinx 1773.

Based on Parkinson's original drawing, this finished watercolour *(opposite)* of *Abutilon indicum* subspecies *albescens* is inscribed 'Fredk Polydore Nodder Pinxt 1778'. The plant grows in tropical parts of Australia and South East Asia, and is commonly known as lantern bush. During the *Endeavour*'s three-week stay at Rio de Janeiro, Banks and Solander collected numerous plants including a fast-growing, evergreen climber now identified as *Stigmaphyllon ciliatum* – Parkinson's painting of which is shown here *(above)*. Once a drawing or painting had been completed by Parkinson, Solander made a note on the back of the name of the species, while Banks noted the locality where it had been found.

Two watercolours based on Parkinson drawings. The painting of *Ficus superba (opposite)* is inscribed 'Fredk Polydore Nodder Pinxt 1782' and was made for Banks. Frederick Polydore Nodder (fl. 1770s–c.1800) is also attributed with painting *Astelia nervosa (above)*, commonly known as bush flax. As usual, Parkinson had made colour notes on the reverse of his sketch and Nodder was able to follow these in completing his painting. The notes for *Ficus superba* read: 'The fruit when young pale green with white specks when older whitish green ting'd wt [with] red wt white specks when ripe dark purple w^{t} white specks'.

RAJA testacea.

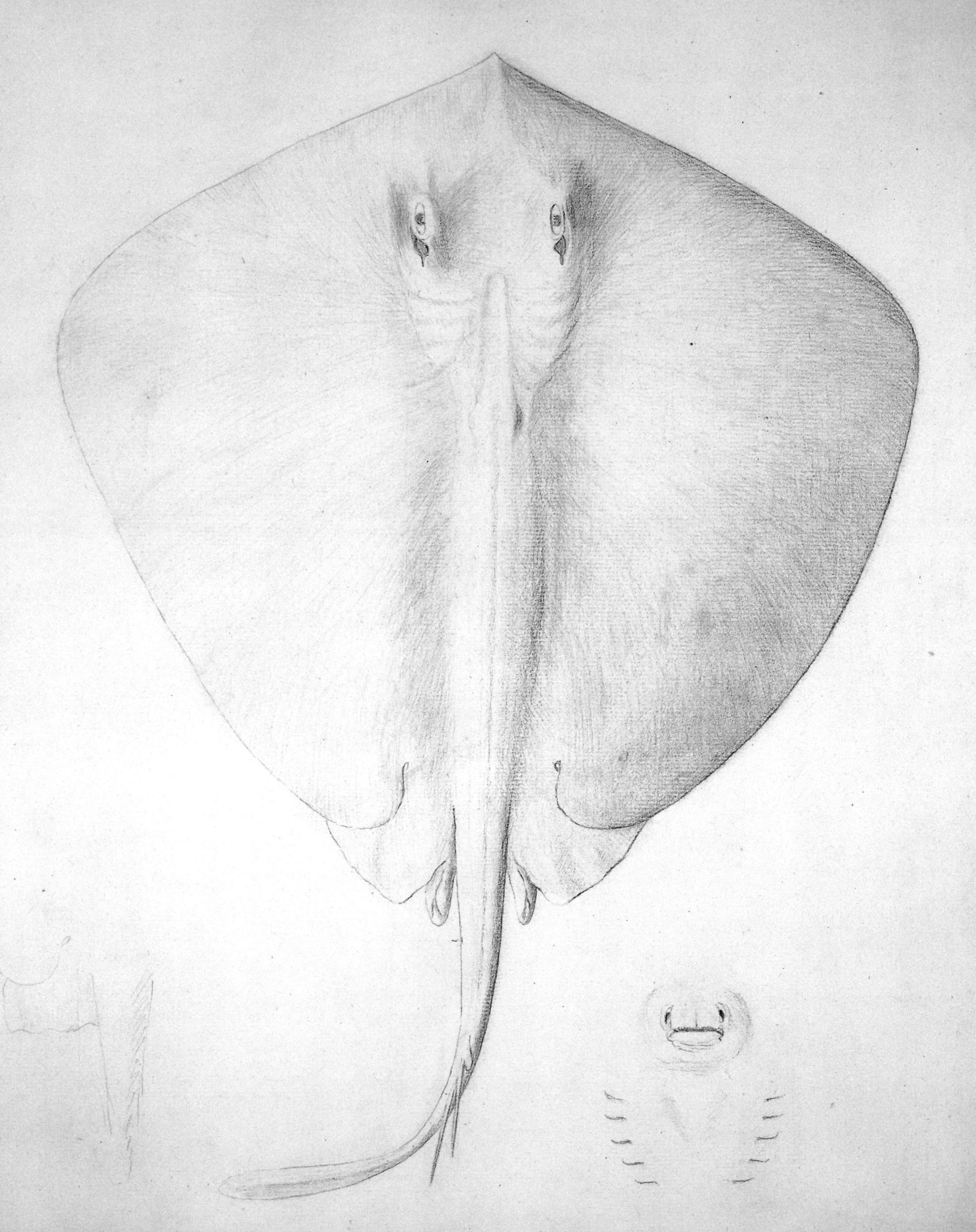

Of all the drawings and paintings made during the *Endeavour* expedition, only four are known to have been made of mammalian species – one of which was the kangaroo *(above)*. This sketch of the animal was made by Parkinson in June 1770 on the Endeavour River near modern-day Cooktown in Queensland. Banks' secretary Herman Spöring drew *Raja testaca (opposite)*, now known as *Urolophus testaceus*, a variety of stingray. The date marked on the drawing is 30 April 1770, giving the location as Botany Bay, which both Banks and Solander referred to as Stingray Bay.

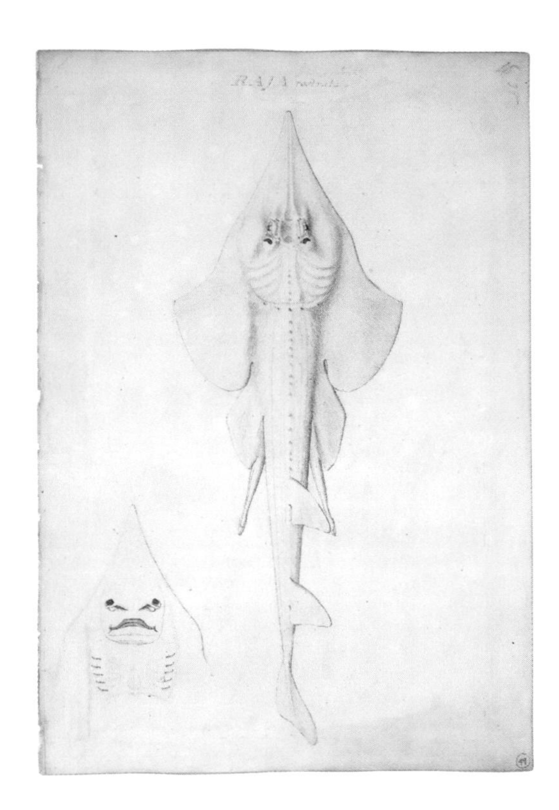

***A scorpion fish** (right)* painted by Parkinson in Madeira in 1768. At this stage of the enterprise, the artist did not feel the pressure that he would be under later, and was able to finish the work he had begun. In New Zealand, the *Endeavour* artists had produced 19 drawings of inshore fish; in Australian waters, however, this number was greatly reduced. Parkinson managed only three, but Herman Spöring was able to make seven pencil studies of sharks, rays and bony fish, of which this is one *(above)*. Labelled *Raja rostrata*, the shark shown here goes by the modern name of *Aptychotrema banksii*. Although there is no record of where the specimen was caught, the date given is 29 April 1770, which indicates that it was found in Botany Bay.

Scorpæna Patriarcha.
Sydney Parkinson pinx.t 1768 –

In April 1769 the expedition reached Tahiti, where Parkinson began painting this species of garfish *(below)*. Solander's name for it was *Esox rostratus*, but the modern scientific name is *Platybelone argala*. The region proved a rich source of marine subjects for the *Endeavour*'s artists: out of the total number of 148 illustrations of fish, 66 were based on specimens found in Tahitian waters. Other sea specimens include the sea slugs *(bottom left & right)* inscribed *Patella* and now known as *Scutus breviculus*. *Chaetodon Gigas (opposite)*, now *Chaetopdipterus faber*, was painted by Parkinson in 1769 from a specimen found at Rio de Janeiro.

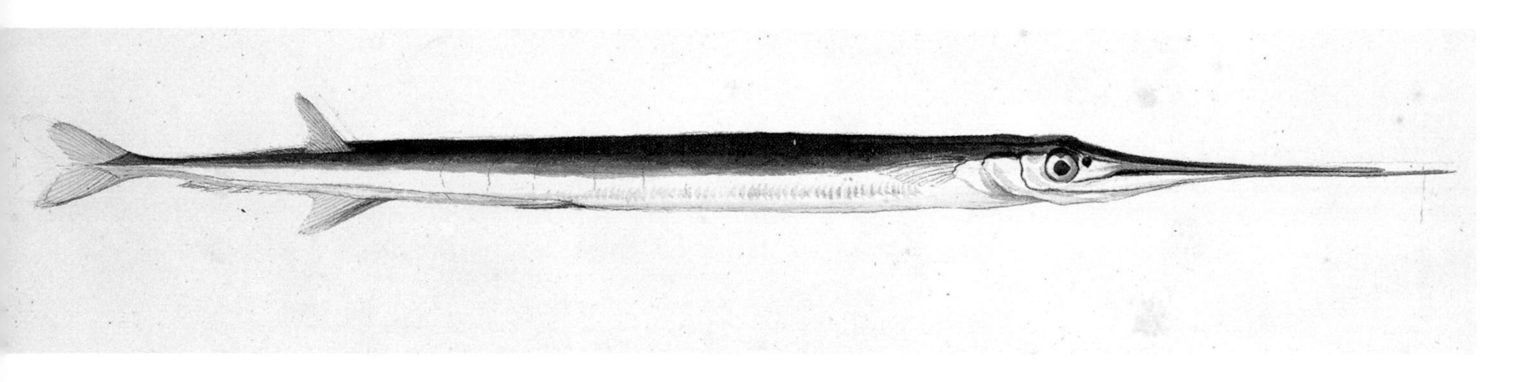

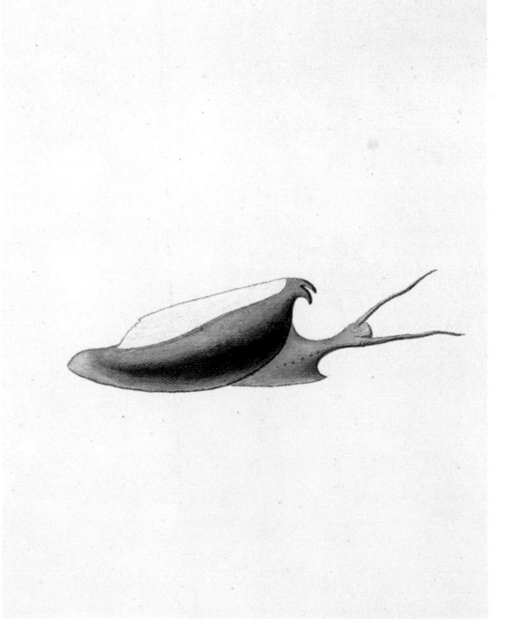

Chætodon Gigas.
Sydney Parkinson pin

Return to the South Seas 1772–1775

James Cook, Johann and George Forster

A brief stop in Tahiti in August 1773 provided the Forsters with time to collect and paint several plants not yet known to botanists. Among these was *Barringtonia speciosa (opposite)* – now renamed *Barringtonia asiatica*.

FOLLOWING THE SUCCESS OF THE *ENDEAVOUR* VOYAGE, both James Cook and Joseph Banks were convinced there should be another expedition to the South Seas to settle once and for all the question of whether or not there was a large southern land mass. But in view of the difficulties experienced during the previous voyage, particularly the perilous passage inside the Great Barrier Reef, Cook wisely felt that two vessels should be involved this time. The Admiralty agreed, and by the end of September 1771, less than three months after the *Endeavour*'s return, the Navy Board was instructed to purchase two suitable vessels. Cook oversaw the purchase and again chose colliers from the same Whitby yard where the *Endeavour* had been built: the *Resolution*, to be commanded by Cook, with a complement of 112, and the *Adventure*, to be commanded by Tobias Furneaux, with a complement of 80.

Cook had hoped to sail in March 1772 but problems with Banks delayed his departure until July. On this trip Banks had proposed to take a team of no fewer than 16 – naturalists, artists, servants, and even two hornplayers. All were to be accommodated on the *Resolution*, which would have to be modified accordingly. Cook was to give up his cabin to Banks and move to new accommodation on an added upper deck. But when the additional accommodation made the ship unseaworthy and had to be removed, Banks threatened to withdraw his people *en masse*. He had employed this technique several times already to get his own way, but this time the Admiralty accepted Banks' withdrawal and quickly appointed its own scientific team, including Johann Reinhold Forster (1729–98) as naturalist, with his 18-year-old son George (1754-94) as assistant and artist.

The elder Forster was born in Poland where he studied theology and natural history at Halle University and then spent 12 years as a country parson near Danzig. During this time he married and raised a family of seven, the eldest being George, and continued his scientific studies. He moved to England in 1766 to teach mineralogy, entomology and other natural history subjects at the highly regarded Dissenters Academy at Warrington. But despite his undoubted talents, he was a difficult character and was sacked in 1769, moving with his family to London where he became well known in scientific circles. This brought him to the notice of the Earl of Sandwich, First Lord of the Admiralty, and so to the post on

191
Barringtonia speciosa. Fl. Austr. p. 47. n. 255.
published in Cook's voyage vol. 1. tab. 24

Pages from the handwritten catalogue *(opposite)* of Johann Forster, in which he named and described specimens collected on the three-year voyage of the *Resolution*. With no more than 290 'stationary' days when the ship was at anchor, time spent on shore was devoted to collecting and drawing as many specimens as possible.

the *Resolution*. Given his temperamental personality, however, he was to prove an irritation to virtually everyone on board and the difficult task of trying to compensate for Johann's shortcomings fell to his son George.

George Forster was extremely gifted, both intellectually and artistically. His father had tutored him in natural history from an early age and he had broadened his education at the Warrington Academy during Johann's period there, but he seems to have had a natural talent for art. During the ensuing voyage he sketched and painted many of the plants and animals seen or collected by his father and himself, particularly those that could not be easily preserved and illustrated later. His plant and zoological drawings from the *Resolution* are preserved in The Natural History Museum in London, having been purchased by Joseph Banks in 1776. Most of the views and landscapes produced during the voyage were actually the work of the *Resolution*'s other official artist, William Hodges (1744–97).

Cook finally left Plymouth on 13 July 1772, intending to sail first to the Cape of Good Hope. From here he was essentially to circumnavigate the globe as close to the South Pole as possible, but this time in an eastward direction to take advantage of the assumed generally westerly winds in these southern high latitudes. He was to investigate any large land masses that he came across in this process, but was to switch his attentions to suitable lower latitude areas during the southern winters. During the first leg of the voyage to the Cape, the two Forsters described and illustrated a variety of marine animals and aquatic birds whose skins began to fill their cabins. The three-week stay at the Cape, and the further five weeks spent there three years later on the way home, gave the Forsters the chance to examine the South African wildlife, some of it in the menagerie in Cape Town, and for George to illustrate some species for the first time. They were overwhelmed by the abundance of undescribed animals and plants around the settlement. So when Johann met Anders Sparrman (1748–1820), a young Swedish doctor, an ex-student of Linnaeus and an accomplished natural historian, he persuaded Cook that Sparrman should join the scientific party on the *Resolution* to assist them.

After leaving South Africa in late November 1772, they continued due south, sighting the first of many icebergs on 10 December. A few days later they were stopped by ice and for a month the ships sailed first westward and then eastward along the edge of the pack but with no sign of land. In mid-January the sea to the south was clear and on the 17th the ships crossed the Antarctic Circle – for the first time in history. Finally, at 67°15'S, solid field ice stopped them, only 120 kilometres (75 miles) from the undiscovered Antarctic continent. Cook then sailed north-east into the southern Indian Ocean, but lost touch with the *Adventure* in thick fog on 8 February. Furneaux had orders to rendezvous with him at Queen Charlotte Sound in New Zealand under such circumstances so the *Resolution* continued north-east to about 45°S then south-east to 62°S by which time she was again surrounded by huge icebergs. On 24 February, Cook decided they could go no further south and for almost a month they sailed eastwards.

Finally, on 17 March 1773, Cook gave up his high latitude search for the season and bore north-east to New Zealand, sighting the South Island on the morning of the 25th. He spent the whole of April in Dusky Sound, a remote stretch of water near the South Island's southern tip which he had discovered during the *Endeavour* voyage. There were ample supplies of fresh water and food to replenish the *Resolution*'s seriously depleted stores, and plenty of opportunities for exploring the Sound's numerous inlets and islands and for meeting the Maoris. By 5 April the Forsters and Sparrman had collected and described 19 birds, three fish and six plants. By the time they left Dusky Sound in May they had collected many more, but in the process the naturalists' conditions on board became less than pleasant. Johann wrote that his cabin 'was a Magazine of all the various kinds of plants, fish, birds, shells, seeds etc. hitherto collected: which made it vastly damp, dirty, crammed, & caused very noxious vapours ...'

Cook made his way along the west coast of the South Island, reaching Queen Charlotte Sound on 18 April to find the *Adventure* awaiting him. There was to be no prolonged and relaxed stay in their 'winter quarters' as Furneaux, and no doubt most of the seamen, had expected and hoped. Instead Cook intended to use the time to sweep yet another unexplored section of the Pacific. From New Zealand he intended to sail east as far as about 135°W at between 41° and 46°S, quite a high latitude for this time of year. He would then turn north and return to New Zealand via Tahiti and various more-or-less unexplored island groups.

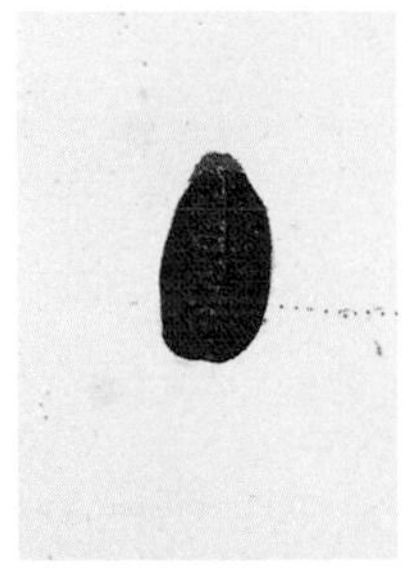

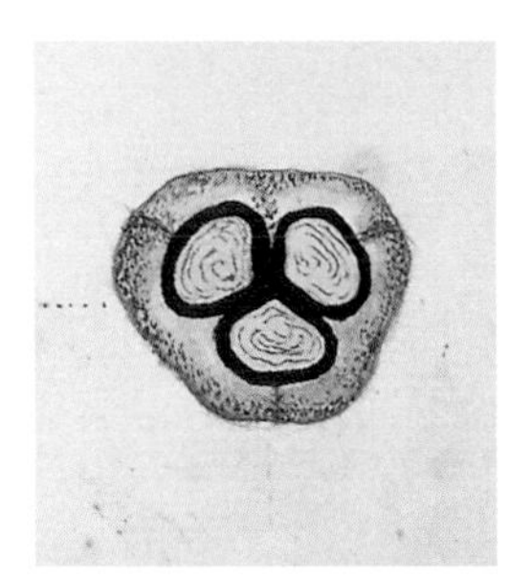

The two ships sailed on 7 June and followed Cook's plan very closely. They actually sailed further east than he had originally intended, reaching 133°30'W without sighting land before turning north towards the Tuamotu Islands, then west to Tahiti. A month later they left with their decks groaning under the weight of new supplies and extensive natural history and ethnographic collections and many of George's drawings. But instead of returning to Queen Charlotte Sound by the shortest route, Cook now sailed west, through the Society Islands and the Friendly Islands before finally leaving Tonga on 8 October. Towards the end of the month, as they approached the western entrance to Cook Strait through which they would have to pass to reach the Sound, they were hit by appalling weather. Cook did not get the *Resolution* safely into her anchorage until 3 November – but there was no sign of the *Adventure*. Furneaux eventually got her into Queen Charlotte Sound, but not until after Cook had already left. Furneaux eventually left New Zealand in late December and returned home via the Cape, arriving in England in July 1774.

In the meantime, Cook made his second great sweep of the Pacific including two more forays south of the Antarctic Circle. Having arrived in Queen Charlotte Sound the *Resolution*'s rigging was repaired and she was thoroughly cleaned, her planking re-caulked and her stores restocked. A note was left in a bottle for Furneaux giving him an outline of Cook's plans including his intention to visit Tahiti, Easter Island or the Society Islands the following winter. After a final search for the *Adventure* around the New Zealand coast, the *Resolution* left alone on 26 November 1773. As they travelled south the weather worsened. Despite the dangers, Cook pressed on, crossing the Antarctic Circle on 20 December. They spent four intensely cold days inside the Antarctic Circle before recrossing it again on Christmas Eve. Christmas Day was calm and, despite the presence of many icebergs, Cook allowed the crew to celebrate it in the usual drunken and ribald way. Johann, though, was depressed, for he saw these long ocean passages as a waste of his time and depriving him of the opportunities to make significant discoveries which Banks and Solander had enjoyed on the previous voyage.

The *Resolution* left the ice in early January 1774 and by the 11th had reached more than two thirds of the way between New Zealand and South America at a latitude of about 48°S. Here Cook changed course towards Cape Horn and raised the hopes of some of the seamen that they were on their way home. But very shortly they turned further to the south and two weeks later crossed the Antarctic Circle yet again. By 30 January, solid field ice and thick fog forced Cook to turn north, convinced that the sea ice reached all the way to the pole.

The ship now left the Antarctic Circle, and the ice, for the last time, sailing first north-east then north in search of land that the navigator Juan Fernandez had supposedly found at about 38°S. Cook found no land and, towards the end of February, decided to give up the search and go directly to Easter Island. But as they started on the new course the Captain was taken dangerously ill with 'bilious colick', probably from a severe gall bladder infection, which put him in bed for almost a week. To provide Cook with fresh food, otherwise totally unavailable, Johann Forster had a dog he had brought from Tahiti killed and cooked. It apparently worked, and to everyone's relief Cook gradually recovered.

After a four-day stopover at Easter Island in mid-March 1774, Cook sailed to the Marquesas, then to Tahiti, and west through the Polynesian islands, before sailing south and west of the Fiji group and arriving at an island group he named the New Hebrides. From late July to the end of August he made a detailed survey of the coastline, before sailing on to the fourth largest island in the Pacific, which he named New Caledonia. Here he spent the month of September, charting its treacherous 480-kilometre-long (300-mile-long) north-east coast. Cook and the botanists were ashore a few days later to examine the strange Cook Pine, *Araucaria columnaris,* which, at 30 metres (100 feet) high, has branches no more than 2 metres (six feet) long. These pines were the source of a good deal of amusement at the expense of the unpopular Johann who, when the crew were viewing them from the ship, had wagered a dozen bottles of wine that they were actually basalt columns. Halfway between New Caledonia and New Zealand, they came across another island with its own curious pine, the Norfolk Island Pine, before finally reaching Queen Charlotte Sound once more on 17 October 1774.

In these two great sweeps through the Pacific islands alone, Cook had achieved an enormous amount – charting, surveying and collecting. But all this was essentially an adjunct to the main purpose of the voyage, the search for the southern continent. To complete this work, Cook now had to traverse the southern Pacific again and search the southern

Two views of **Medusa pelagica** *(opposite),* a species of jellyfish. The title *Medusa* derives from the ancient Greek gorgon of the same name, a mythical creature whose head was covered with writhing, biting snakes – a reference to the jellyfish's hair-like, stinging tentacles – while *pelagica* denotes the open sea.

Atlantic before returning home. But this last attempt was to be in vain. 'I can be bold to say that no man will ever venture farther than I have done and the lands which may lie to the South will never be explored,' he wrote on 27 January 1775. 'Thick fog, snow storms, Intense cold and every other thing that can render Navigation dangerous one has to encounter, and these difficulties are greatly heightened by the inexpressible horrid aspect of the Country, a Country doomed by Nature never once to feel the warmth of the Sun's rays, but to lie for ever buried under everlasting snow and ice.'

Cook finally reached England on 30 July 1775. Once home, he busied himself preparing the results of his second voyage for publication, and in the process became embroiled in a major row with Johann Forster. Forster claimed that it had been agreed he should write up the public account of the voyage. This was hardly likely, but early in 1776 a compromise was reached under which Cook and Forster would publish in collaboration. By the summer even this arrangement had broken down. Eventually, three separate accounts of the voyage were published. With the full support of the Admiralty, Cook's two-volume work appeared in May 1777, including 12 charts and 51 monochrome engravings of places, people and artefacts mainly based on originals by Hodges. But six weeks earlier George Forster had published a two-volume account prepared with his father, while Johann's own single-volume observations on '... Physical Geography, Natural History and Ethnic Philosophy', appeared the following year. With no money or time to undertake expensive engravings, and lacking official support, neither of the Forster publications contained illustrations.

Despite the title of Johann's book, it contained very little detailed information about the natural history collections and, as in the case of Cook's first voyage, very little about them was published until much later. Many of the plants and animals collected were totally 'new' at the time of the *Resolution* voyage, but Johann's descriptions of the specimens, many of them referring to George's illustrations, were not published until 1844, 46 years after Johann's death. By this time most of them had been described and named by other authors. The surviving botanical and zoological collections, and the illustrations from the *Resolution*, are justifiably revered and treasured for their historical, scientific and artistic value; but at the time they did not receive the attention, and certainly not the publicity, that they deserved.

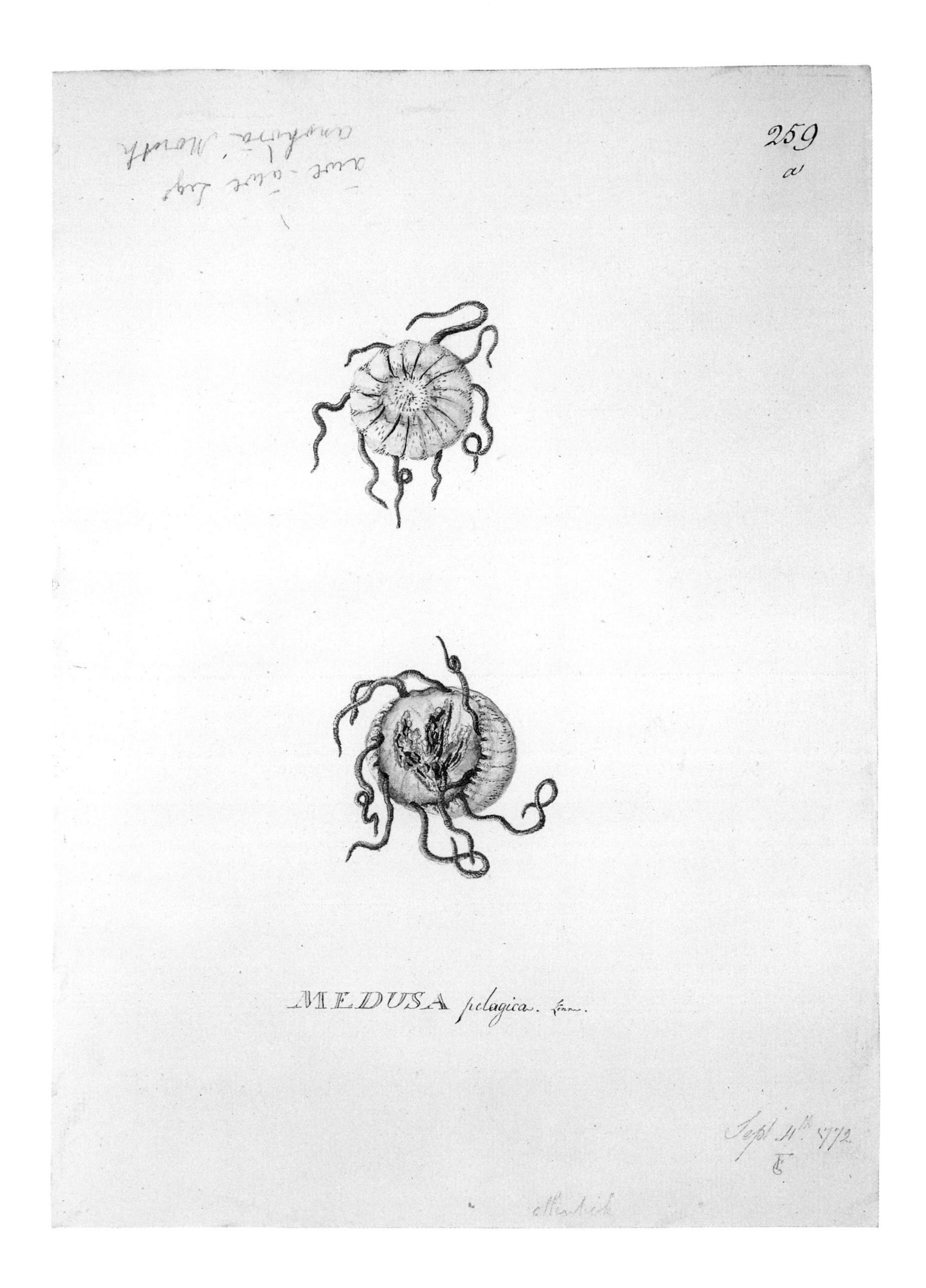
259
a
MEDUSA pelagica. Linn.
Sept. 11th 1772

The **Resolution** ***made two month-long visits*** to the Cape of Good Hope, during which these pencil sketches were made by George Forster. Neither George nor his father had time to travel beyond the Cape and so relied on drawing from bought specimens and living animals in the Cape Menagerie. The white-tailed gnu or black wildebeest, *Connochaetes gnou (opposite),* is a large antelope found only in South Africa. The giraffe sketch *(above)* was actually copied by George from an oil painting owned by the Governor of the Cape, Joachim van Plettenberg.

Bos Connochaetes. Mas.
Antilope Gnu S.N. XIII. 189.2.25

In his notes, Johann Forster decribes seeing an eland *(right)*, a type of South African antelope, at the Cape Menagerie in 1772. Forster knew the animal as *Antilope oryx*, but its modern name is *Taurotragus oryx* – *tauro* from the Latin *taurus* for 'bull'. The signed and completed illustration *(above)* of the kaffir rail or *Rallus caerulescens*, formerly *Rallus cafer* – is another of the species which the Forsters encountered in the Cape Menagerie. In the wild, it lives in swamps and reedbeds, a habitat which George Forster seems to have attempted to reproduce here.

30.

George Forster painted a snow petrel *(above)* – *Pagodroma nivea* – on 30 December 1772 in the 'Southern Ice Ocean'. Johann Forster's notes state that snow petrels were to be seen south of latitude 52°, particularly in the vicinity of ice. In compensation for the freezing conditions this far south, the naturalists on board the *Resolution* had the opportunity to see and study the rich – and then almost undocumented – bird life of the region which, as well as this petrel, included the fulmar *(opposite)*, the snow prion, the Antarctic prion, and the emperor and king penguins.

S.L. 36°
16.

81.
36 inches high
Aptenodytes patachonica J. R. Forster in Comm. Gotting. 3. p. 137.
published by Mr Pennant in his genera of birds tab. 14.

Johann Forster had already seen king penguins on the *Resolution*'s first foray into sub-Antarctic waters back in December 1772, but the size of the specimens he saw on South Georgia *(opposite)* in 1775 impressed both him and Cook. There were so many on South Georgia and neighbouring islands that they blanketed the snow in a carpet of black. Also spotted there were 'gregarious' seals, which Forster labelled *Phoca antarctica (below)*. This was the last time Cook and his crew would observe either animal, for Cook had finally abandoned his search for the elusive Southern Continent.

***Notes to the painting of a ray** (opposite), Raja edentula*, now known as *Aetobatus nariniri*, date it to 10 May 1774, during the *Resolution*'s second visit to Tahiti. This was just one of numerous natural history studies that Johann Forster made, with the help of his son's artistry, in Tahiti. The island's flora and fauna were not the only subjects for Forster's scrutiny, though – he observed the people, too, making anthropological studies of the inhabitants of Tahiti and the other islands of the South Pacific, and becoming something of a pioneer in this new field of research. Other marine specimens include *Tetrodon hispidus (top)*, now known as *Arothron meleagris*, a type of puffer fish which was collected from Raietea, one of the Society Islands. The fish inflates itself when threatened in order to elevate its poisonous spines. *Blennius superciliosus (above)*, now known as *Clinus superciliosus,* was fished from waters around the Cape of Good Hope on the *Resolution*'s return journey to England.

250.
Raja edentula.
Tahiti May 10th 1774.

Unfinished paintings of *Prosopeia tabuensis (above left)* – the 'red shining-parrot' from the Friendly Islands – and *Ninox novaeseelandiae*, commonly known as the mopoke or morepork *(above right)*. The mopoke was just one of the bird specimens taken from Queen Charlotte Sound by the industrious Forsters and Sparrman, whose other discoveries included the New Zealand plover, the piopio and kokako, several parakeets, the spotted cormorant, and the weka, a type of flightless rail. Several of the species they recorded would be extinct a century later. *Halcyon leucocephala acteon*, or the grey-headed kingfisher *(opposite)*, is from the Cape Verde islands where Cook stopped *en route* to the Cape of Good Hope. This George Forster painting is one of the few from the voyage in which the subject is set in a finished landscape.

60.
Alcedo senegalensis β. S.N. XIII: 456.

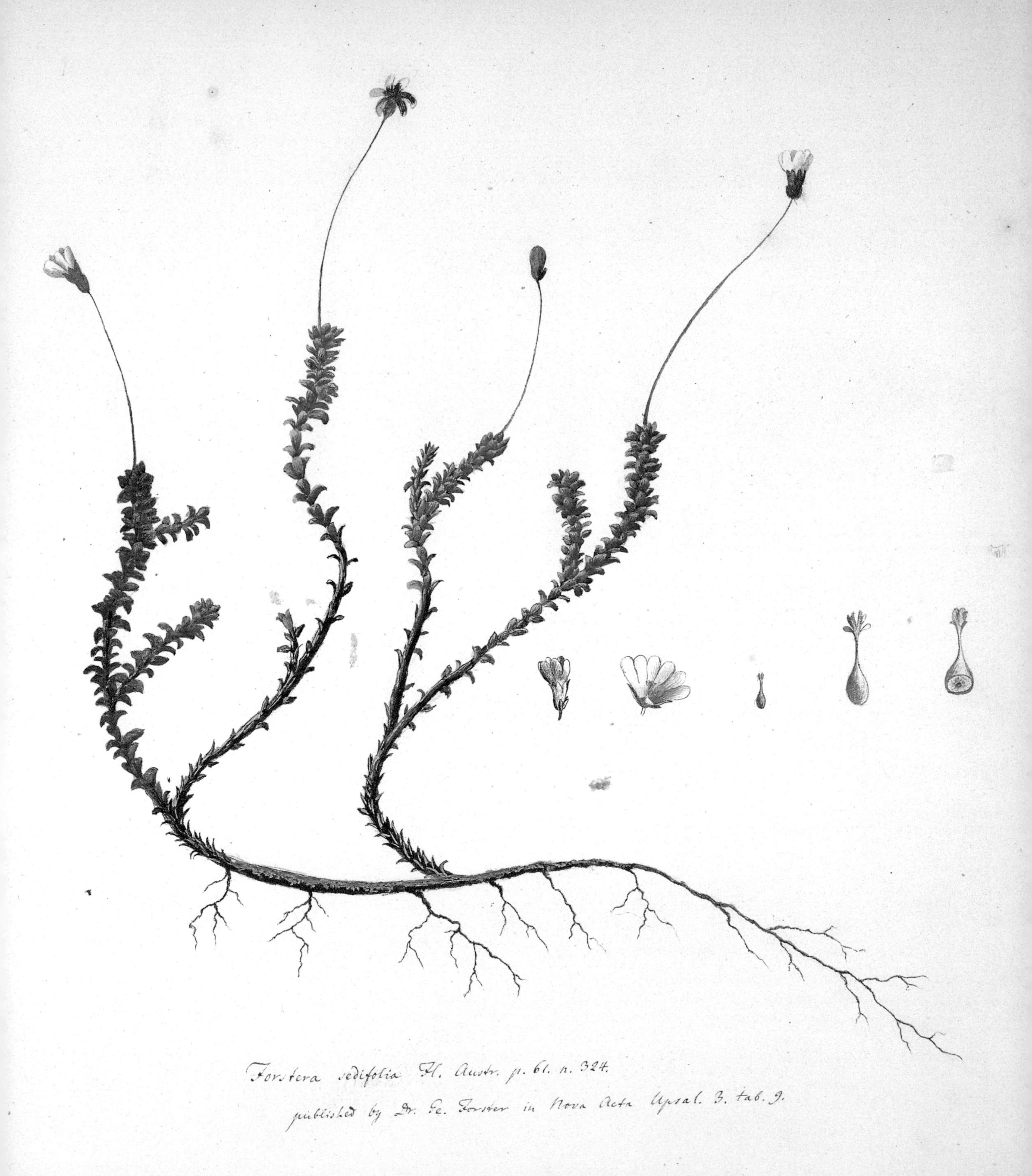
Forstera sedifolia Fl. Austr. p. 61. n. 324.
published by Dr. Ge. Forster in Nova Acta Upsal. 3. tab. 9.

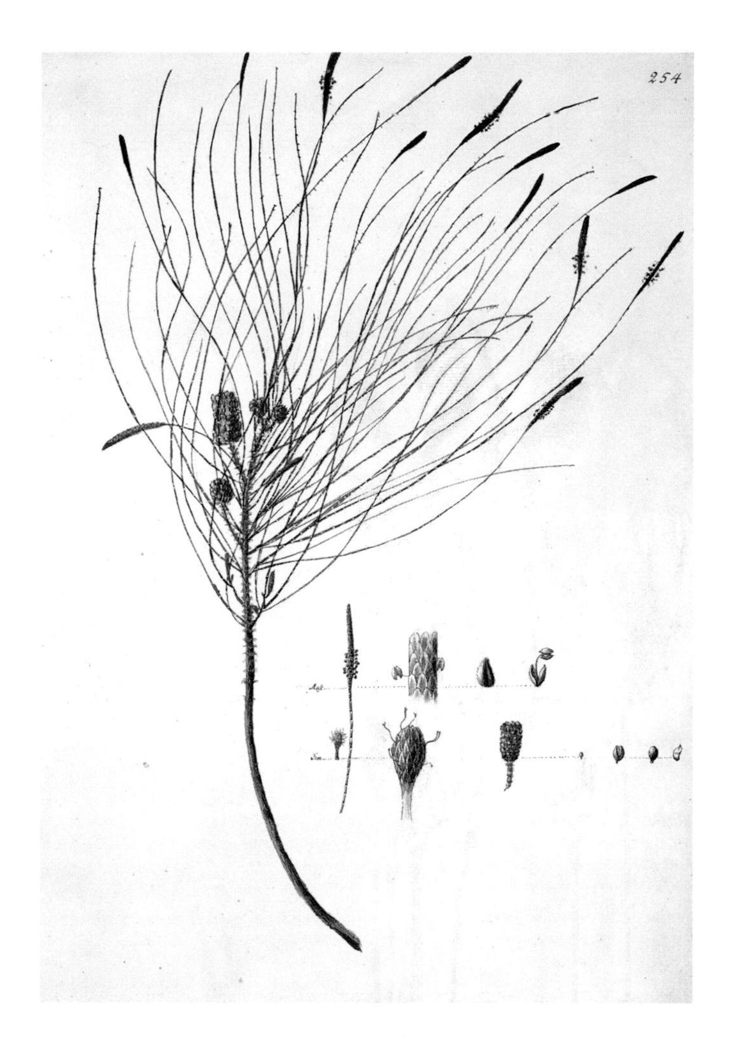

Forstera sedifolia *(opposite)* was found at Dusky Bay (Dusky Sound), the *Resolution*'s first port of call on her initial visit to New Zealand. This was just one of a large collection of native flora and fauna specimens that the Forsters and Sparrman recorded, greatly increasing the list formerly compiled by Sir Joseph Banks and the botanist Daniel Solander as a result of Cook's previous visit to New Zealand. *Casuarina equisetifolia (above),* with details of bark, seedheads and seeds, was found in Tahiti. The species belongs to a genus of trees known commonly as beefwood or she-oak, which have scale-like leaves. The species name – *equisetifolia* – alludes to the tree's similarity in appearance to horsetails, *equisetum* (from the Latin *equus*, or 'horse').

248
Passiflora aurantia Fl. Austr. p.62. n. 326.

A finished watercolour of a species of passion flower, *Passiflora aurantia (opposite)* is dated 8 September, and gives the locality as New Caledonia – the year would therefore have been 1773. An unfinished and undated drawing of a plant found at Queen Charlotte Sound, New Zealand, is inscribed *Trichilia spectabilis (above)*, but the plant's modern name is *Dysoxylum spectabile*. The *Resolution* first visited the Sound in May 1773, returning there again over a year later in October 1774. The naturalists were able to collect specimens on both occasions.

Two of some 300 botanical illustrations made on a voyage that would eventually cover 113,000 kilometres (70,000 miles). *Melaleuca astuosa (above)* has since been renamed *Metrosideros collina*. Although in close-up this plant from Tahiti bears some resemblance to *Melaleuca*, botanical illustrations can be misleading for they often give no idea of scale. While some *Melaleuca* species reach heights of up to 12 metres (40 feet), *Metrosideros* can attain a towering 25 metres (80 feet) – a difference which cannot be conveyed by individual twigs from the different genera. A twig named as *Jatropha gynandra (opposite)*, with seedheads and seeds, and signed 'G F' by George Forster, is more likely to be *Jatropha curcas*.

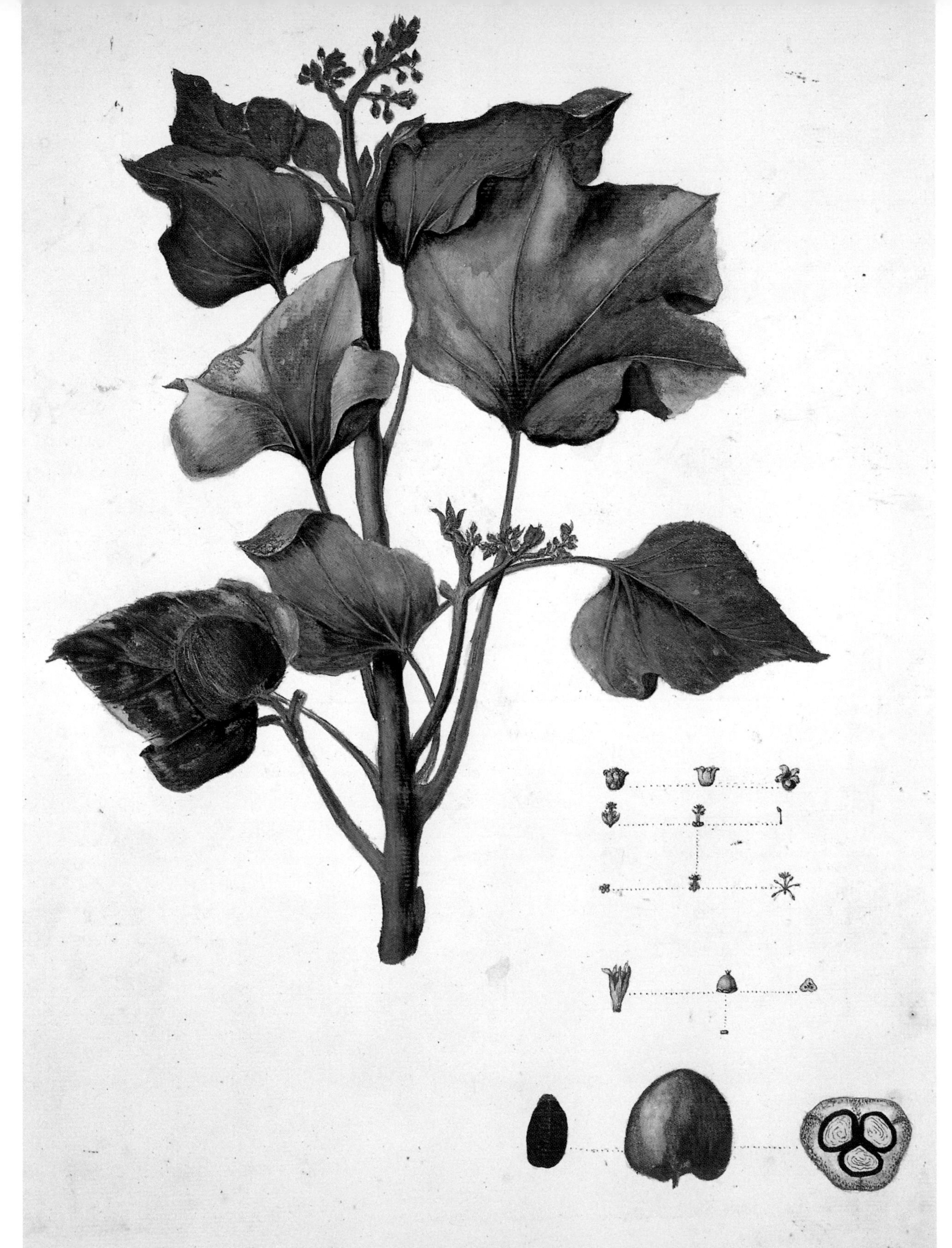
JATROPHA gynandra.
St. Jago
Curcas? Fl. Atlant. in Commentat. Gotting. 9. p. 70. n. 148.

***The thistle-like* Carthamus lanatus** *(opposite)* was sketched in August 1772 at Madeira, but was not completed until February or March of 1773, as were a number of other plant drawings including *Convolvulus digitalis (above left)*. A note at the bottom suggests that the species may actually be *C. althaeoides* as it has the typical trumpet-shaped flowers of the genus. Originally given the name *Leptospermum callistemon (above right)*, this plant from Queen Charlotte Sound is now known as *Metrosideros scandens*.

Sketched Aug. 1. 1772.
CARTHAMUS lanatus. LINN.
Gotting. 9. p. 66. n. 131.

Charting Australia 1801–1805

Matthew Flinders & Ferdinand Bauer

The small helmet orchid *Corybas unguiculatus (opposite)* grows up to 3 centimetres (1 inch) high and is native to Australia. The specimen drawn here by Ferdinand Bauer was probably collected at Woolloomooloo in 1804.

In November 1778, Joseph Banks was elected President of the Royal Society, the most senior scientific post in Britain, and one which he was to occupy until his death in 1820. One of the most significant recommendations he was to make within the first few years of his appointment was that Australia's Botany Bay be developed as a convict outpost, as the end of the American War of Independence in 1781 had put a stop to the practice of shipping criminals to North America. His suggestion was eventually adopted and in May 1787 the First Fleet of eleven ships sailed from Spithead with 443 sailors and 800 convicts to found the new penal colony. They arrived at Botany Bay in January 1788, but Captain Arthur Phillip, the Fleet's commander and colony's first Governor, was unimpressed and decided to settle instead at Port Jackson, the future Sydney, a few kilometres to the north.

During the first few years of the colony's existence little effort was devoted to surveying the coastline. Consequently, by 1795 when Matthew Flinders, a 21-year-old midshipman, arrived at Sydney on HMS *Reliance* carrying the colony's new Governor, John Hunter, only about 160 kilometres (100 miles) of the coast to the north and south had been surveyed. For the next five years Flinders extended the surveys with the limited facilities available locally and, when he returned to England in August 1800, recommended to Banks that an official naval expedition should be sent to Australia to build on the charting work of Cook almost 30 years previously and on the natural history studies begun by Banks and Solander.

Banks agreed to the proposal and so did the Admiralty, particularly when it got wind of a major French expedition to Australia in the vessels *Géographe* and *Naturaliste,* despatched in October 1800 under the command of Nicolas Baudin. Britain was, of course, not the only nation interested in Australasia and the Pacific. The French, with whom Britain had been at war since 1793, also had their eyes on colonising possibilities in the region, especially since it was just possible that Australia, or New Holland as it was usually called, was not a single land mass but was bisected by a north-south passage from the Great Australian Bight to somewhere in the region of the Gulf of Carpentaria. So both warring nations were to have exploratory expeditions in Australian waters at the same time, each guaranteeing the other immunity from hostile military or naval interference. And both were to make important geographic and scientific discoveries.

Matthew Flinders' published account of his expedition on board the *Investigator,* entitled *A Voyage to Terra Australis,* featured plates of the topography along the south coast of Australia, including the entrance to Port Phillip *(below)* and the mountains at Spencer's Gulph (Gulf) *(opposite).*

But all this was in the future. By the end of November 1800, the 30-metre-long (100-foot-long) sloop *Xenophon* had been selected for the expedition, to be renamed *Investigator* and placed under the command of Flinders, now a Lieutenant. The ship was to carry a naval complement of 80 and a civilian staff comprising an official landscape painter, William Westall (1781–1850), an astronomer, John Crosley, mineralogist John Allen (b. 1775), a naturalist and a natural history painter. To serve as expedition naturalist, Banks finally chose Robert Brown (1773–1858), a 27-year-old Scot from Montrose who had studied medicine at Edinburgh and, since 1795, had been an ensign and assistant surgeon in the Fifeshire regiment of fencibles. After serving for three years in northern Ireland, Brown had been transferred to London. He was recommended to Banks, after Banks' first choice, Mungo Park, withdrew. Brown was to be assisted on the *Investigator* by a gardener from Kew Gardens, Peter Good (d. 1803), and an Austrian artist, Ferdinand Bauer (1760–1826), now widely considered to have been one of the greatest natural history artists that ever lived.

Bauer was born in Feldsberg, then in lower Austria but incorporated into the new state of Czechoslovakia after World War I. His father was the official painter to the court of the reigning Prince of Lichtenstein, but died when Ferdinand was only a year old. Ferdinand inherited his father's artistic talent and, from an early age, developed a particular interest and skill in drawing plants. He obtained employment initially in Feldsberg and later at the University of Vienna under Nicolaus von Jacquin (1727–1817), Professor of Botany and Director of the University's Botanical Garden. Here, in 1784, he met John Sibthorp (1758–96), Sherardian Professor of Botany at Oxford, who invited Bauer to participate in a planned botanising visit to Greece. After 18 months in Greece, Bauer returned with Sibthorp to England at the end of 1787 together with more than 1,500 sketches made on the trip. Settling in Oxford, Bauer worked for several years converting these sketches into finished paintings which were used to illustrate Sibthorp's *Flora Graeca*, published between 1806 and 1840 (long after Bauer's death) and considered by the renowned British botanist Joseph Hooker to be 'the greatest botanical work that has ever appeared'. Through this work, Bauer became known to Banks who accordingly recommended him for the *Investigator* voyage.

The *Investigator* sailed from Spithead in July 1801, arriving at Cape Leeuwin in south-western Australia in December. Flinders' first task was to investigate and survey the southern coast, but before doing so the ship spent a couple of weeks in King George Sound, giving Brown and his team the opportunity to study the area. Within a few days they had collected 500 plant specimens, almost all of them belonging to new species, and a number of animals – the beginnings of a vast collection which Bauer assiduously sketched. On his survey of the coast, Flinders scheduled shore excursions in the Archipelago of Recherche and made a running survey of the Great Australian Bight before calling at Port Lincoln near to where eight of the crew were lost in a cutter despatched to find water. After surveying and collecting in Spencer Gulf and Kangaroo Island, the *Investigator* met up with the *Géographe* in Encounter Bay on 8 April 1802 as Baudin was surveying the coast in the opposite direction. A month later they reached Sydney, mooring in Sydney Cove.

It was now 10 months since leaving England and the ship was in urgent need of repairs, so Flinders stayed at Sydney until 21 July to complete them. It was an eventful port call. Brown and Bauer were able to make a number of important collecting excursions, including one to the Blue Mountains, and Brown wrote to inform Banks that they had already obtained some 300 new plant species and that Bauer had made more than 500 sketches. But there were also important political developments. The second ship of the French expedition, the *Naturaliste*, was already at Sydney when Flinders arrived, while the *Géographe* limped in with a sickly crew a few weeks later. Both expeditions were relieved to know that their respective countries were no longer at war, though the peace was to be short-lived.

With the ship temporarily repaired and the crew rested, Flinders now set out on his next great objective, to complete the circumnavigation of the continent and to check for evidence of a major waterway into the landmass from the north coast. The *Investigator* sailed northwards, successfully navigating the dangerous waters of the Great Barrier Reef faced by James Cook 30 years earlier. After several shore excursions on the mainland and on the offshore islands, and a number of amicable encounters with the native Australians, they rounded Cape York and entered the Gulf of Carpentaria in early November 1802. For the next four months Flinders surveyed the coastline of the Gulf and of Arnhem Land to the west,

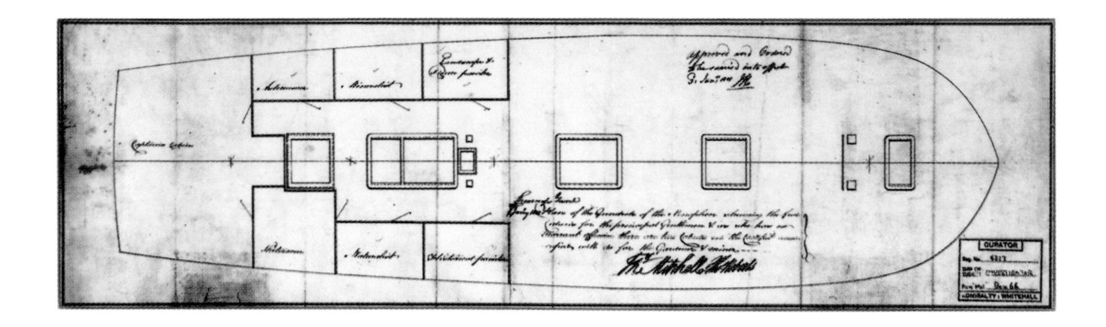

but found no evidence of a significant passage to the south; Australia was a single land mass. Several encounters with the native Australians in this region were less than cordial, at least one episode ending in the death of one of them. There was also an unpleasant contact with the local botany when Brown, Bauer and the gardener Good were made violently sick by eating the nut of the cycad tree, an experience they had had with the same plant a year previously on the south coast but appear to have forgotten. Despite the effect on Bauer, who seems to have been particularly ill, it did not prevent him from carefully illustrating this plant or the hundreds of others collected in this botanically and zoologically fruitful region.

The ship had been showing signs of serious deterioration for some time and by March 1803 Flinders had decided that they would have to cut the detailed surveys short and return to Sydney as quickly as possible, though still completing the circumnavigation. Contrary winds forced them across the Timor Sea to Kupang Bay, Timor, where they were able to get fresh supplies. But the two-month voyage back to Sydney was difficult. Many of the crew were by now exhausted and sick. Dysentery broke out and there were two deaths during the passage and a further four, including that of Good, shortly after they reached Sydney on 9 June. The *Investigator* was by now in such bad condition that she could not continue the survey work.

It was decided that Flinders and many of the expedition personnel should return to England and hopefully bring out a replacement ship to complete the original objectives. But Flinders' bad luck continued. His first attempt ended when the new ship he was in was wrecked on a coral reef on 17 August 1803, only a week after they had sailed. Flinders struggled back to Sydney in the ship's cutter and set sail again on 20 September with a three-vessel rescue fleet, including his own vessel, the tiny 29-tonne *Cumberland*, to continue from the wreck site to England with some of the survivors while the rest would be taken back to Sydney. But after picking up the survivors things went from bad to worse. The *Cumberland* was a poor sailing vessel, and sustained considerable damage as she rounded northern Australia and headed into the Indian Ocean via Timor. She was not going to make the Cape of Good Hope so Flinders decided to head for the French possession of Ile de France (Mauritius) instead. Unfortunately, he had not heard that hostilities between England and

Part of a blueprint of HMS **Investigator** *(opposite),* which was large enough to carry a complement of more than 80. This section of the vessel was given over to the Captain's cabin plus cabins for the 'principal Gentlemen', as the plan notes describe them, being the astronomer, mineralogist, landscape painter, natural history painter and naturalist.

France had broken out once more, this time in the Napoleonic Wars that would last for more than a decade. As Flinders was not travelling on the ship listed on his passport, the local Governor, Decaen, arrested him as a suspected spy. Despite protests from England and even from France, he remained a prisoner for six-and-a-half years until Decaen was recalled and Flinders could finally return home in October 1810.

In the meantime, the work of the *Investigator*'s surviving naturalists continued. Brown, Bauer and the mineralogist Allen had decided to stay behind in Sydney when Flinders left for England. They would await his return for up to 18 months and then, if he had not come, would make their own way home. They made a series of short joint collecting trips in the Sydney region until November 1803 when Brown decided to visit Van Dieman's Land (Tasmania). He intended to be away for only 10 weeks, but the trip extended to nine months during which he made extensive collections and a number of new discoveries, including that of the Tasmanian wolf or thylacine, now thought to have become extinct in the 1930s. Brown did not, of course, have the advantage of an artist with him because Bauer, still based in Sydney, was making his own trips, including one as far as the settlement of Newcastle some 150 kilometres (93 miles) to the north. Bauer had no idea when Brown would come back, but by the middle of 1804 had decided that he had exhausted the natural history of the Sydney region and that he should look further afield.

By now the decrepit *Investigator* was being refitted to prepare her for her return to England, together with Bauer, Brown and their collections. Bauer decided to travel the 1,000 kilometres (620 miles) to Norfolk Island from where he expected to be picked up a few weeks later by the *Investigator*. After leaving Sydney in August 1804, only days before Brown returned, Bauer spent almost eight months on the tiny island before the unreliable *Investigator* collected him. During this time Bauer combed Norfolk Island collecting and illustrating the animals and particularly the plants. The resulting drawings formed the basis of Stephan Endlicher's 1833 publication *Prodromus Florae Norfolkicae* listing 152 Norfolk Island plants, several named in honour of Bauer. Eventually, in late February 1805, the *Investigator* collected Bauer and returned him to Sydney to rejoin Brown and Allen. They left for England on 23 May with some 36 large cases

Matthew Flinders *(opposite)* finally returned to England in October 1810 and immediately began work on the official narrative of his voyage. He died in July 1814, the year his account was finally published.

containing their collections and Bauer's drawings – together with a live wombat. After a tedious but uneventful journey, the *Investigator* limped into Liverpool on 13 October 1805, four years and three months after she had left.

Brown's collection contained almost 4,000 plant species, some 1,700 of them new to science, together with about 150 bird skins, a variety of other vertebrate and invertebrate animals and an extensive mineralogical collection. Bauer had produced more than 2,000 sketches, about 1,750 of plants and the rest of animals. The whole lot was transferred to Sir Joseph Banks' house in Soho Square in London where, for the next 10 years, Brown worked on the collections, preparing accounts for publication while Bauer produced hundreds of finished paintings and drawings from his detailed sketches with their copious colour notes. Most of these paintings were officially the property of the Admiralty who relinquished ownership to the British Museum in 1843.

Brown published only one of a projected series of volumes on Australasian botany, selling only 26 of the 250 copies printed. Bauer intended to publish a series of his plates to accompany Brown's descriptions, but could not find anyone capable of adequately engraving and colouring his superb illustrations. He had to do it all himself, and sold only a few copies of the 15 plates that he published. He would have been amazed to know that a copy sold in 1982 for $25,000. But both Brown and Bauer did see some of their work published successfully in their lifetimes. Despite Flinders' ordeal, he began the preparation of his official account, *A Voyage to Terra Australis,* as soon as he returned to London. It finally appeared in 1814, including an 80-page botanical appendix by Brown and illustrated by 10 of Bauer's botanical plates.

In 1814 the disillusioned Bauer returned to his native Austria, together with his extensive collections, most of his colour-coded *Investigator* sketches and some finished paintings. Here he continued life as a botanical illustrator, dying in 1826 in Hietzing, now a suburb of Vienna. As a result, the sketches are now mainly held in the Naturhistorisches Museum in Vienna. But the superb finished paintings found their way to the British Museum and then to The Natural History Museum in London where they form one of the best collections of original zoological, and particularly botanical, illustrations anywhere in the world.

CAPTAIN FLINDERS. R.N.

Autograph Copy of Parole on his release from six years Captivity in the Isle of Mauritius.

I undersigned, captain in His Britannic Majesty's navy, having obtained leave of His Excellency the captain-general to return in my country by the way of Bengal, Promise on my word of honour not to act in any service which might be considered as directly or indirectly hostile to France or its Allies, during the course of the present war

Port Napoleon, Isle de France, 7th June 1810

(Signed) Matthw Flinders

Published by HARGROVE SAUNDERS 24 Tichborne St LONDON W

LITHO PHOTOGRAPHIC INST. 492, NEW OXFORD STREET

Bauer's paintings from the *Investigator* trip are remarkable in their detail, given that there was relatively little time on land to execute them. Bauer developed a unique technique for coping with this problem. Unlike Sydney Parkinson, who when travelling with the *Endeavour* would make a partially coloured sketch from which he later produced the final painting, Bauer devoted his time on location to making very detailed pencil sketches and colour coding them according to his own complex system. On return to London he used these coded sketches to make paintings capturing the subtlest nuance in colour. Superb examples of the finished results include the yellow daisy burr, *Calotis lappulacea (opposite);* the cabbage palm, *Linistona humilis (above);* and *Velleia pubescens (overleaf).*

Leptoryhnchos scaber *(right),* found at King George Sound on the coast of Western Australia. It was one of hundreds of plant specimens collected during the *Investigator*'s short stop there. Writing to his brother Franz on 22 May 1802 from Sydney Cove, Bauer recalled: 'After a journey of five weeks from the Cape (of Good Hope) we got to see the land of New Holland for the first time on 7 December 1801 after which, on the 8th, we anchored in King Georges Sound, Western Australia and remained there till the 4 January 1802 in the course of which we made a number of short land excursions during which we found many new plants.'

Among the hundreds of plants amassed during the short stay at King George Sound was *Cephalotus follicularis (left),* collected on 1 January 1802. Commonly known as the Albany pitcher plant, it is summer flowering. The flowers rise on long, slim stems above a cluster of the treacherous pitchers, in which trapped insects are dissolved by the plant's digestive fluid.

Ottelia ovalifolia *(above left),* the swamp lily, was found by Robert Brown in November 1803 at Parramatta, Hawkesbury and Richmond in New South Wales. Three months before at Shoalwater Bay, Queensland, Brown collected a fruiting specimen *(above right),* later named *Mackinlaya macrosciadea* after explorer John McKinley (1819–72), who famously set out to rescue Robert O'Hara Burke (1820–61) and William Wills (1834–61), the first Europeans to cross Australia from south to north. A species from the genus named after Joseph Banks is *Banksia coccinea (opposite),* the latter term from the Latin *coccineus*, or scarlet-coloured. It is commonly known as the Scarlet Banksia and is native to Western Australia.

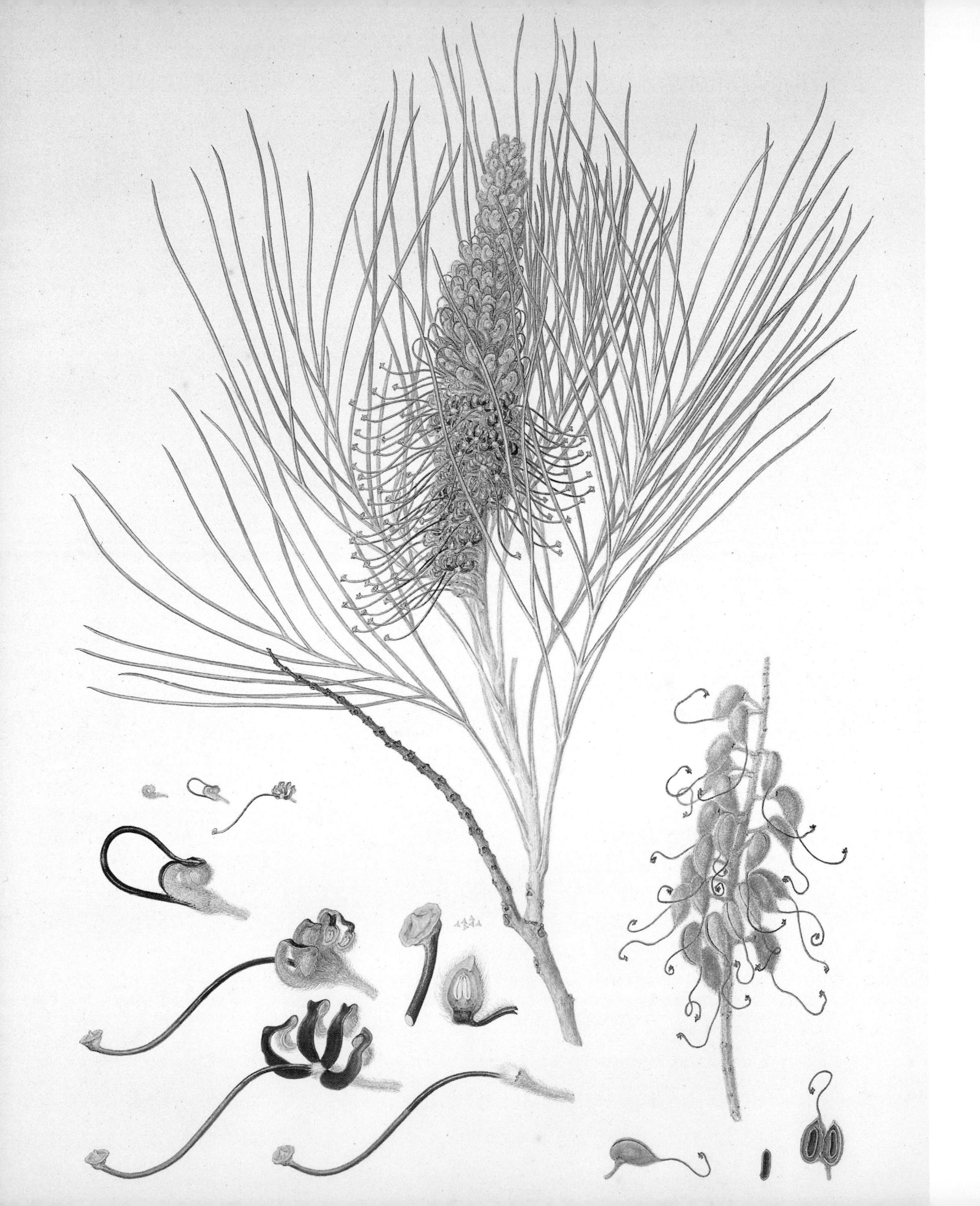

The yellow flower of *Grevillea pteridifolia (opposite)*. Robert Brown established the genus, which includes a large number of plants, naming this species *chrysodendron*. A year earlier, however, in 1809, Joseph Knight had entitled it *pteridifolia*, and according to the international standard of nomenclature, this name had to remain. In Queensland *G. pteridifolia* is commonly known as the golden parrot tree because its mass of yellow flowers attracts lorikeets and other birds. *Lomandra hastilis (left)*, a species of mat rush, grows on the opposite side of the country, in Western Australia. This suggests that Bauer's specimen was found some time in December 1801 or early in January 1802, when the *Investigator* expedition was based at King George Sound.

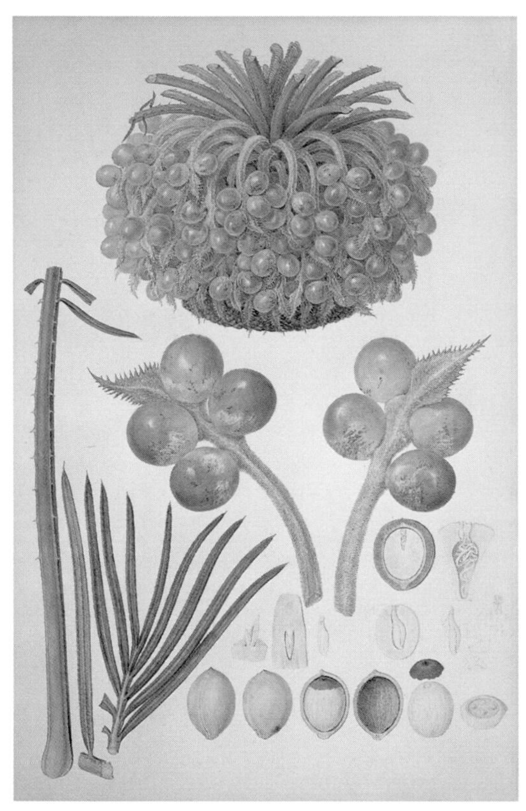

The male cycad tree, *Cycas media,* with its male cone *(above left)*, nuts, highly toxic seeds, leaf and stem sections *(above right)* and the exposed female cone *(opposite)*. All species in the genus *Cycas* are dioecious, with pollen-bearing male cones and seed-bearing female ones borne on separate plants. Captain Cook's crew had already encountered the tree on their earlier survey of Australia in 1770. Joseph Banks recorded that: '... some of our people who tho forewarned ... eat one or 2 of [the seeds and] were violently affected by them both upwards and downwards ...' Despite the seeds' toxicity, the Aboriginals used them as a foodstuff, grinding them into flour and sago – but only after soaking and cooking.

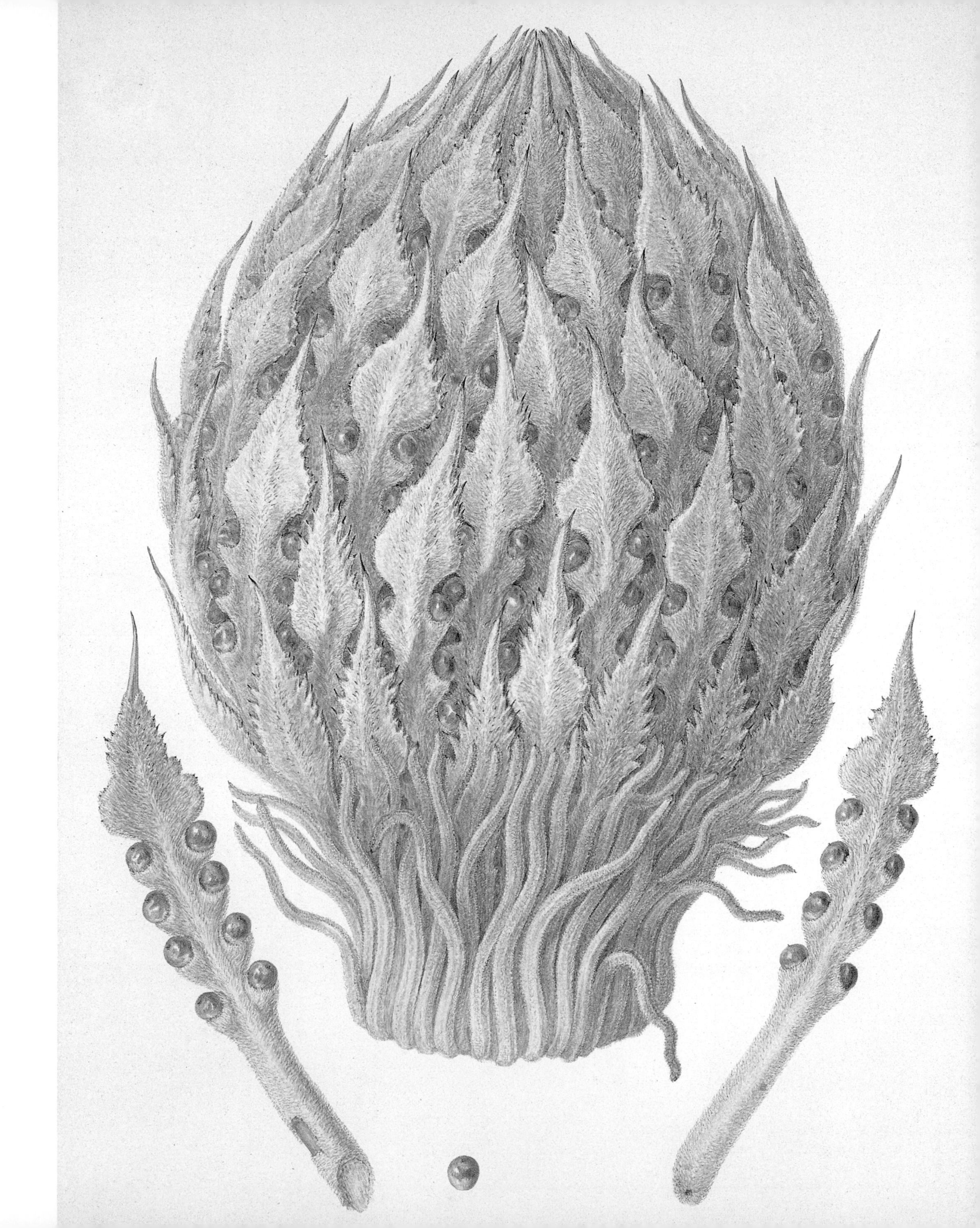

From the Torres Strait Islands off the northern coast of Queensland came what is thought to be the kapok tree, *Cochlospermum gillivraei*, of which Bauer painted the pods *(above)*. Flinders noted in his journal that on the islands '... A species of silk-cotton plant was plentiful; the fibres in the pod are strong, and have a fine gloss, and might perhaps be advantageously employed in manufacture ...' This particular tree is different from other kapok trees, such as *Ceiba pentandra*, the silk-cotton tree, which grows in tropical America and Africa and from which kapok is produced for use as mattress stuffing, and in sound insulation. The roots of the Australian kapok are a traditional food of the Aboriginal people. Bauer's precision artistry is evident in both the kapok pods and his illustrations of a palm leaf *(opposite)*.

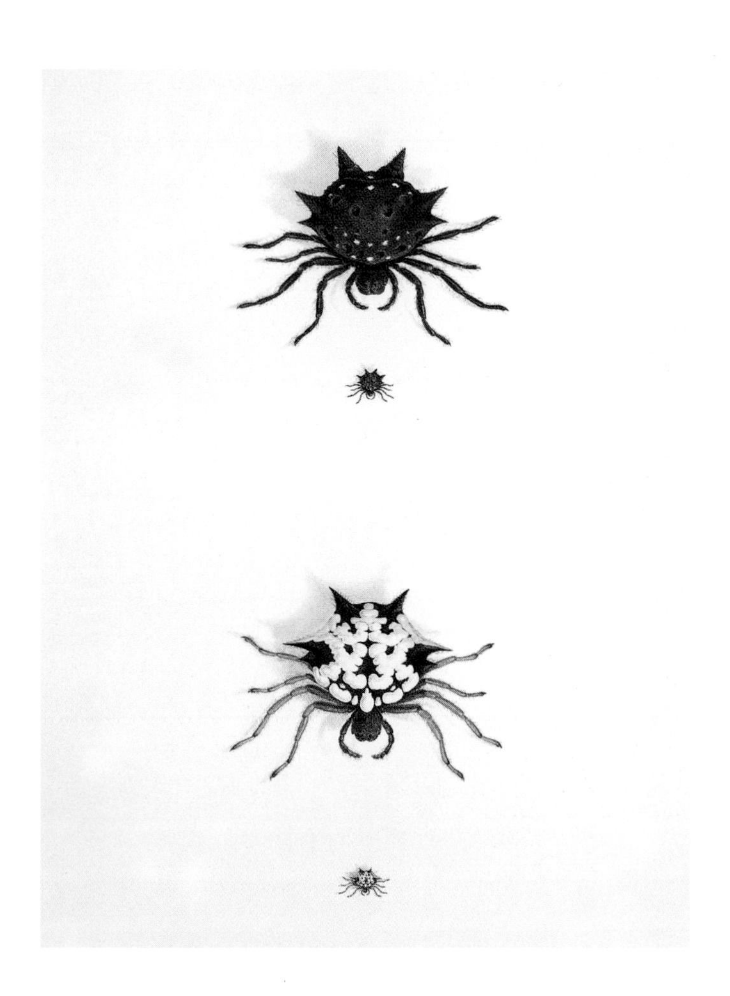

The weedy or common seadragon *(left), Phyllopteryx taeniolatus,* was probably dredged up in a mixed haul out of King George Sound. Flinders' journal for 3 January 1802 relates that a variety of small fish were brought up. '... These were of little use, as food; but with the shells, sea weeds, and corals, they furnished amusement and occupation to the naturalist [presumably Brown] and draughtsman [Bauer], and a pretty kind of hippocampus [seahorse], which was not scarce, was generally admired ...' Also common in Australian waters was the spider crab, *Gasteracantha mimax (above).*

Variously called the blue, blue swimming or blue manna crab, or the sand or sandy crab, this is *Portunus pelagicus (above)*, named after Portunus, the Roman god of harbours, and the most common edible crab. The complex colour-coded numbering system used by Bauer, in which every shade of colour was designated a numerical code of up to four figures, meant that he was able to draw the specimen while it was still fresh and before its natural colouring faded *(opposite)*. Later, when he began the watercolour version, he could reproduce the specimen's original tones with the greatest accuracy.

N° XVIII
N° 245

The black-footed rock wallaby *(opposite)* or *Petrogale pencillata*, painted by Bauer from a specimen Matthew Flinders described as a '... small kangaroo of a species different from any I had before seen ...' Brown has a similar response to the discovery of a skink *(below), Egernia cunninghami.* He records in his diary for 22 December 1801, at King George Sound: 'They found plenty of Aptynidotes miner & of a species of Lizard different from what I had before seen ...'

Acanthaluteres brownii *(opposite below)* is named after Robert Brown, and one of its common names is Brown's leather jacket. Bauer's drawing provided the sole information for the original description of the species. An illustration of another leather jacket *(opposite above)* also demonstrates the continued value of his work. The fish was named *Brachaluteres baueri* in 1846, no connection being made between it and *Balistes jacksoniaunus*, which had been described and named earlier. In 1985, a study of Bauer's drawing revealed that these two names in fact referred to the same species. It is still known as *Brachaluteres baueri*. Another specimen painted by Bauer *(below)* belongs to the genus *Pterois*. It is believed to have been caught at Strong Tide Passage, Queensland, on 28 August 1802.

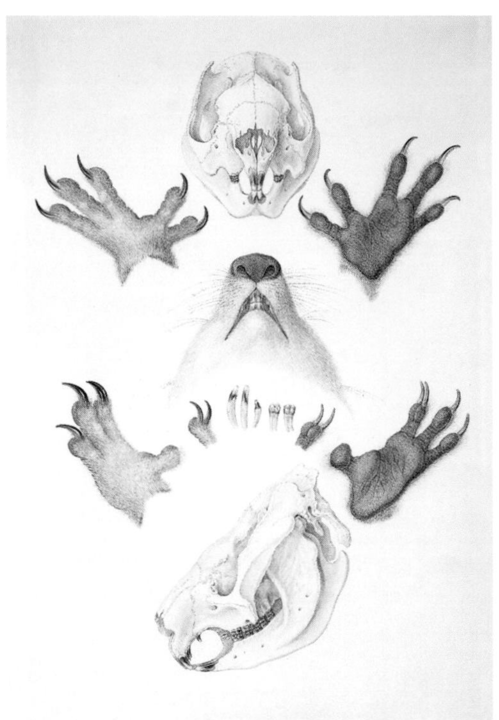

The first Europeans to see a wombat, *Vombatus ursinus (above left)*, were probably sailors shipwrecked on Preservation Island in the Bass Strait in 1797. Like the Aboriginal inhabitants, the sailors regularly ate wombat. They were rescued by a party including Flinders, who captured a live wombat and presented it to Governor Hunter in Sydney. It died six weeks later, however, and its body was sent to Joseph Banks in London. Banks also hoped to receive a pair of live koalas, *Phascolarctos cinereus (above right)*, which Robert Brown proposed to send him, but in the end it was thought unfeasible because of the animals' restricted diet. Banks questioned the very existence of the platypus, *Ornithorhynchus anatinus (opposite)*, but in 1800 Governor King was able to send him a preserved specimen. Two years later, Bauer drew this specimen at Port Jackson, New South Wales.

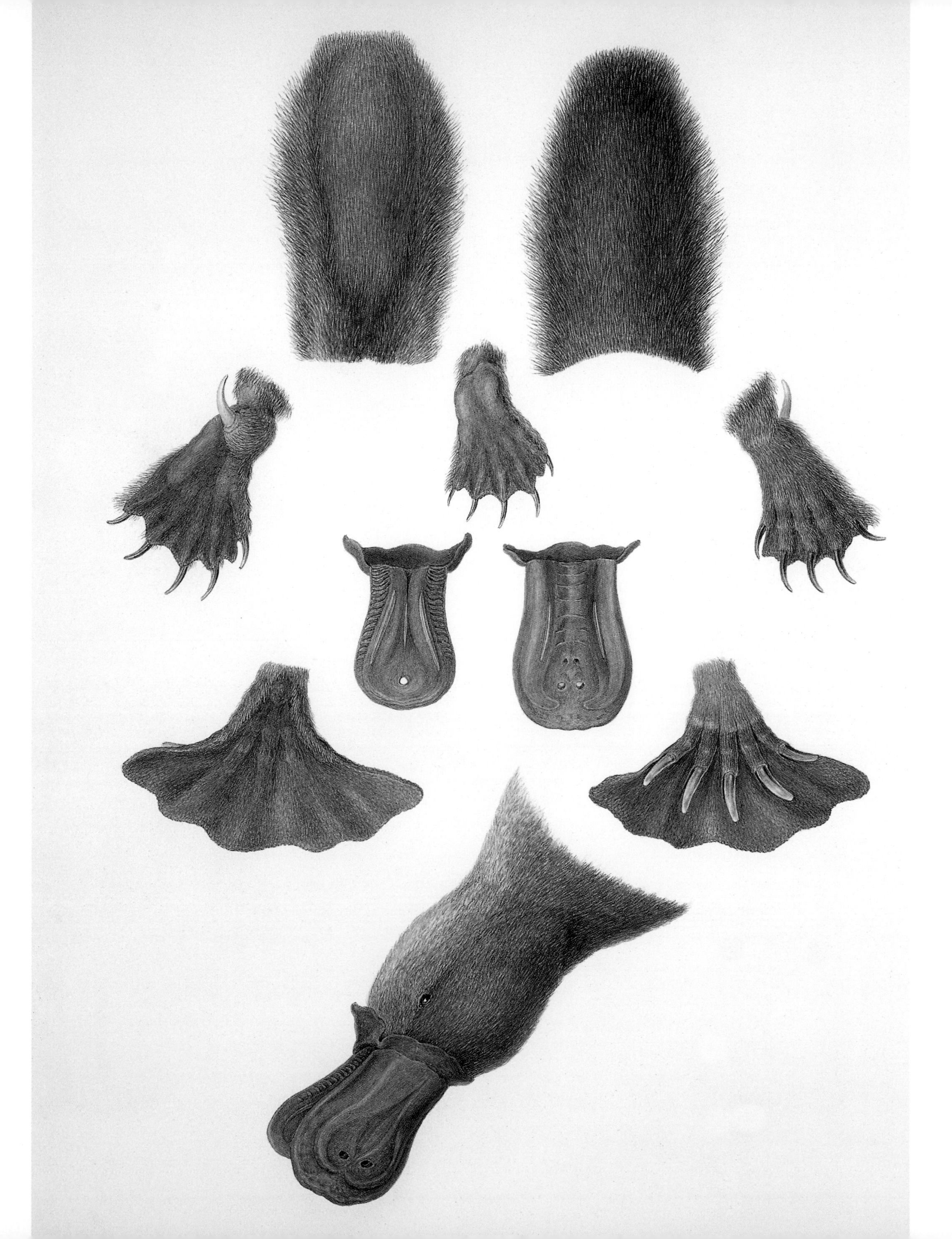

Sailing with the *Beagle* 1831–1836

Charles Darwin

Darwin saw this bird *(opposite)*, originally named *Tanagra darwinii* but now called *Thraupis bonariensis*, feeding on cactus at Maldonado in Brazil. His interest in bird life predated the study of Galapagos finches.

Charles Darwin's five years spent circumnavigating the world in HMS *Beagle* under the command of Robert FitzRoy is one of the most famous journeys ever undertaken. His extensive collections and his observations filling numerous notebooks during the voyage ultimately led him to his theory of evolution by natural selection set out in his world shaking book, *On the Origin of Species*. But the *Origin* was not published until 1859. In December 1831, when the *Beagle* sailed from Plymouth, Darwin was simply a 22-year-old, upper middle class, unsuccessful Edinburgh medical student who had just obtained a rather mediocre divinity degree from Cambridge and expected to become an Anglican parson in a quiet English country parish. Although he was interested in natural history, passionately so in beetles, and had some influential scientific acquaintances, young Charles showed little sign that he was to become one of the greatest biologists that has ever lived.

Robert FitzRoy, on the other hand, seemed clearly destined for a distinguished career. From an aristocratic family descended from Charles II, FitzRoy had been an academically outstanding student at the Royal Naval College at Portsmouth which he had entered as a 13-year-old in 1818. He had been promoted to Lieutenant at the early age of 19 and had been in command of the *Beagle* since he was 23, when he was transferred to her at Rio de Janeiro in 1828 to replace her previous Captain who had committed suicide, his own sad fate some 40 years later.

After two years surveying in South American waters, including charting the whole of the coast of Tierra del Fuego, FitzRoy had returned to England in October 1830 with four Fuegians who he intended to educate in Britain and then return to their homeland so that their communities could benefit from this contact with 'civilisation'. Although one of them had died from smallpox, and the Admiralty was not too keen on the idea, FitzRoy's opportunity to fulfil his plan came when he was reappointed to the *Beagle* to continue the South American survey and return to England via the Pacific and the Cape of Good Hope.

Based on his earlier experiences, FitzRoy proposed to the hydrographer 'that some well educated and scientific person should be sought who would willingly share such accommodations as I could offer, in order to profit by the

Tanagra Darwini.

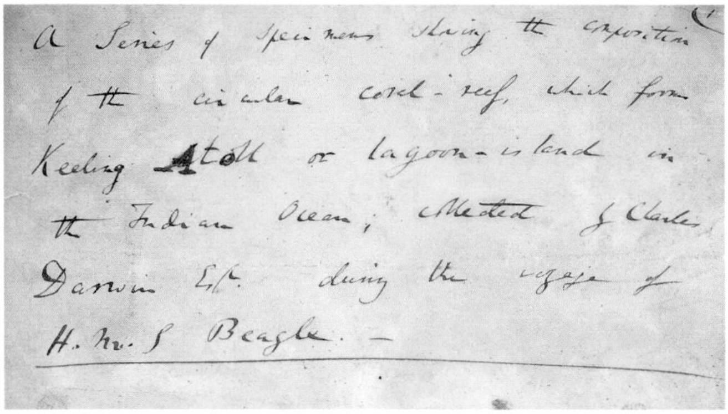
A Series of Specimens showing the composition of the circular coral-reef, which form Keeling Atoll or lagoon-island in the Indian Ocean; collected by Charles Darwin Esq. during the voyage of H.M.S Beagle. –

opportunity of visiting distant countries yet little known.' On the recommendation of his scientific friends, the encouragement of most of his family and overcoming limited resistance from his father, Charles Darwin became that person, accommodated and fed by the navy but otherwise at his own, or rather his father's, expense. Apart from his duties as naturalist, Darwin was to eat with FitzRoy and to some extent alleviate the Captain's inevitable isolation. Because of his easy-going nature he was able to act as something of a buffer between the crew and the highly respected, but often irascible and moody, FitzRoy.

The *Beagle*, technically a brig-sloop, was tiny, only 27 metres (90 feet) long and 8 metres (25 feet) in the beam. She had to accommodate 74 people including the naval personnel, Darwin, the three Fuegians and the by-now more or less obligatory official artist. The presence of a professional artist was particularly important because, although several of the naval complement, including FitzRoy, were quite accomplished draughtsmen, Darwin had virtually no artistic talent.

The first official artist was Augustus Earle, 38 years old at the beginning of the voyage and one of the oldest men aboard. Engaged by FitzRoy, though like Darwin victualled by the Admiralty, Earle was a good choice because he had already spent 20 years recording the scenery in remote parts of the world including South America, Australia and New Zealand. But he was also an excellent portraitist. During the first year of the voyage, spent off the Atlantic seaboard of South America, Earle produced a series of landscapes and watercolours of shipboard life, as well as illustrating material collected by Darwin both at sea and ashore. Darwin had ample time ashore during this period, including several weeks in Rio when he and Earle shared a house, and subsequently during port calls at Montevideo and Bahia Blanca. But Earle's health was not good and he had to stay behind in Montevideo when the *Beagle* sailed for her first visit to Tierra del Fuego in December 1832. Because of his absence, the depictions of the Fuegian natives, including those by FitzRoy (and by Earle's replacement, Conrad Martens, during the second visit), are not as good as they might have been. Earle was still in Bahia when the *Beagle* returned in late April 1833, but he was not fit enough to rejoin her and eventually made his own way back to London where he died in 1838.

Notes made by Darwin *(opposite)* on a collection of coral specimens he collected during a study of reefs in 1836 at the Cocos Islands (formerly the Keeling Islands) in the Indian Ocean. Darwin realised that the coral reefs were the end result of a very long process involving the gradual subsidence of ancient volcanic islands and the simultaneous build-up of the coral structures on their slopes. He later published his famous study *The Structure and Distribution of Coral Reefs* in 1842. At the time of its publication, his theory on coral formation was refuted, but it has since been fully confirmed by deep-sea research.

In the meantime, Darwin had undertaken two of his longest overland journeys, the 960 kilometres (600 miles) from El Carmen to Buenos Aires and subsequently a further 960 kilometres (600 miles) up the Parana River to Santa Fé and back, all without the benefit of a professional artist. But by the time he rejoined the *Beagle* in Montevideo, having had to pass through a revolutionary blockade of Buenos Aires to do so, a replacement had been found. Conrad Martens, a 32-year-old landscape painter, had heard of FitzRoy's need for an artist when he reached Rio de Janeiro in the summer of 1833 on his way to India. He immediately went to Montevideo to offer his services to FitzRoy and was promptly engaged, sailing with the ship when she left Montevideo for the last time in December 1833. Nine months later, when the *Beagle* reached Valparaiso, Martens was put ashore because there was no longer room for him. But during this relatively short period, including the *Beagle*'s second visit to Tierra del Fuego when FitzRoy's protégés were found to have reverted to their original state, and the passage around Cape Horn and up the Pacific coast of South America, Martens was extremely prolific. As a result, his pencil drawings and many watercolours form much the greatest part of the pictorial record of the voyage. Moreover, after he left the *Beagle* in Valparaiso, Martens visited and sketched in Tahiti, the Bay of Islands in New Zealand and Sydney, all places that the *Beagle* also called at on her way home. Some of these non-*Beagle* illustrations were subsequently used by FitzRoy in his official narrative of the voyage.

From Valparaiso the *Beagle* continued northwards, spending another year surveying the South American coast while Darwin made further fascinating, and sometimes perilous, land excursions, all the time collecting, observing and recording. Finally, with the surveys completed, the ship left South America in September 1835 heading for the Galapagos Islands as the first stage of her homeward journey. It would take her more than a year to reach England; apart from the Galapagos she stopped at Tahiti, New Zealand, Australia, the Cocos Islands, Cape Town, St Helena and Ascension before a return visit to Bahia and thence to Falmouth where Darwin disembarked on 2 October 1836. For the ship's crew the return journey was a relatively unpressured affair. But Darwin continued assiduously to fill his jars and bottles with specimens and his notebooks with observations.

***The HMS* Beagle,** *(opposite)* under sail in a painting by Owen Stanley dated 1841. Darwin had come on board as a 'gentleman's companion' to the commander Captain FitzRoy, who feared the loneliness of his voyage to chart the coast of South America might drive him mad. The ship's name has since become synonymous with Darwin and the voyage that led to his theory of evolution. Ironically though, the devout FitzRoy believed that Darwin's dissension from the biblical story of creation was tantamount to heresy.

On the approach to the Galapagos, Darwin had already seen ample evidence to convince him that the earth and its inhabitants had undergone massive changes in the past: mountain ranges had been thrust upwards and later been eroded away; major sea level changes had exposed areas which had previously been part of the seafloor; and whole islands and landmasses had appeared and disappeared. While all this was happening, the fossil record indicated that while some animal and plant groups seemed to stay more or less the same through vast periods of geological time, others showed major changes, some becoming totally extinct and being replaced by new ones. And everywhere the living organisms he observed showed wonderful adaptations to exploit the vagaries of their often harsh environments. None of these ideas, even the evolution of animals and plants was novel, but they put Darwin in conflict with FitzRoy whose view of the earth was a strictly biblical one; according to this view, all of the earth's creatures were God's immutable creation, while geological upheavals and mass extinctions had to be related to biblical events such as the Flood.

It was one thing to believe in evolution. But it was quite another to prove it and, more importantly, to explain it. It would take another 20 years of painstaking work before Darwin was ready to go public with his theory of natural selection, but the five weeks that the *Beagle* spent in the Galapagos provided crucial information. The isolated island group consists of a dozen or so tiny islets. As he wandered around them, Darwin was struck by two remarkable phenomena. First, they had a unique fauna; from the extraordinary giant tortoises and marine and terrestrial iguanas to the small invertebrates, many of the Galapagos species were found nowhere else on earth. Though many of them were clearly related to similar forms on the South American mainland, the Galapagos species were subtly, sometimes dramatically, different. 'It was most striking,' Darwin wrote, 'to be surrounded by new birds, new reptiles, new shells, new insects, new plants, and yet by innumerable trifling details of structure, and even by the tones of voice and plumage of birds, to have the temperate plains of ... Patagonia, or the dry deserts of northern Chile, vividly brought before my eyes.' Even more remarkably, several of the islands had their own unique forms, distinct from those on neighbouring islands only a few tens of kilometres away. This was true even for the tortoises, but was even more obvious in other groups.

These curious features were particularly well illustrated by the birds, and especially by what ultimately turned out to be 13 distinct species of small, rather dull-looking birds, all vaguely similar, but with very distinct beaks which seemed to allow them to specialise in different food sources ranging from nuts and seeds, through insects to fruits and flowers. These were what later became famous as 'Darwin's finches', although young Charles did not realise at the time just how important they were. In the absence of an official artist, the Galapagos fauna was not adequately illustrated at the time. The artist who eventually dealt with the islands' birds, John Gould, probably never even saw the *Beagle*, let alone sailed on her; yet his contribution to the overall *Beagle* story was enormous. Unlike Darwin, Gould was an expert ornithologist and, again unlike Darwin, he recognised almost immmediately the significance of the Galapagos finches.

By the time the *Beagle* returned to England, Darwin had long since realised that the life of a clergyman was not for him. Fortunately, his financial security meant that he had no need to earn a living. Instead he threw himself into preparing the *Beagle* zoological results for publication and, more importantly, into the long-term development of his evolutionary ideas. But he obviously couldn't deal with all the animal groups himself and needed to enlist the help of experts. For this reason, in January 1837, he took his collection of mammals and birds to the recently established Zoological Society of London where Gould was the official ornithologist.

Born in 1804, the son of a gardener at Windsor Castle, Gould had received little formal education and started his working life following in his father's footsteps as a trainee gardener at Windsor. But as part of his training he was introduced to the art of taxidermy, a not uncommon skill amongst gardeners at the time, and by the age of 20 had established his own taxidermy business in London. With his Windsor connections, he stuffed several birds for George IV and, in 1827, secured the position of 'Curator and Preserver' for the newly established Zoological Society of London. Six years later he had become superintendent of the Society's ornithological department and had already started out on his lifelong passion for publishing lavishly illustrated, and highly successful, bird books. Though Gould was quite talented as an artist himself, most of the illustrations in his many publications, even if attributed to him, were apparently produced by other

***From Darwin's* Journal of Researches** *(opposite)*, four species of Galapagos finch with different beaks. Darwin questioned the view that each species was created individually by God, and was unchanging: 'As many more individuals of each species are born than can possibly survive,' he wrote, 'and as consequently there is a frequently recurring struggle for existence, it follows that any being, if it vary however slightly in any manner profitable to itself ... will have a better chance of surviving, and thus be naturally selected .... This preservation of favourable individual differences and variations, and the destruction of those which are injurious, I have called Natural Selection, or the Survival of the Fittest.' The finch specimens were carried back to London in the hold of the *Beagle (overleaf)*.

artists, principally his wife Eliza, but also Edward Lear, more famous these days for his nonsense poetry. Gould's own main strengths were as a publishing entrepreneur and as an extremely good ornithologist, and it was in this latter capacity that he made his important contribution to Darwin's evolutionary ideas.

Gould examined all of Darwin's birds and named and described the many new species in the collection. He was very excited by all of these, but particularly by the rather nondescript little birds from the Galapagos. Because of their distinct life-styles, Darwin had assumed that they included representatives of several different groups: wrens, finches, 'gross-beaks' and blackbird relatives. But within six days of receiving them, Gould was able to announce that despite their totally different beaks they were all, in fact, finches. Interesting though this was, the real significance did not at the time immediately strike anyone, including Gould and Darwin. In fact, Gould never claimed any credit for elucidating the theory of natural selection in which the evidence of the Galapagos finches played such an important part.

Gradually, Darwin realised that the finches he had collected on the *Beagle* voyage represented the results of a remarkable 'natural experiment' in evolution. Sometime in the distant past, when the Galapagos were largely uninhabited by birds, some typical finches, presumably from the South American mainland, had managed to reach the islands. Freed from competition from other small specialised feeding birds, these ancestral finches had evolved and speciated to occupy all the available 'small bird niches' and in the process acquired a whole range of un-finchlike beaks – a perfect, if very unusual, example of adaptation by natural selection. Darwin would undoubtedly have reached this conclusion eventually unaided. But the sowing of the first germs of the idea were certainly the work of John Gould, publisher and bird-man.

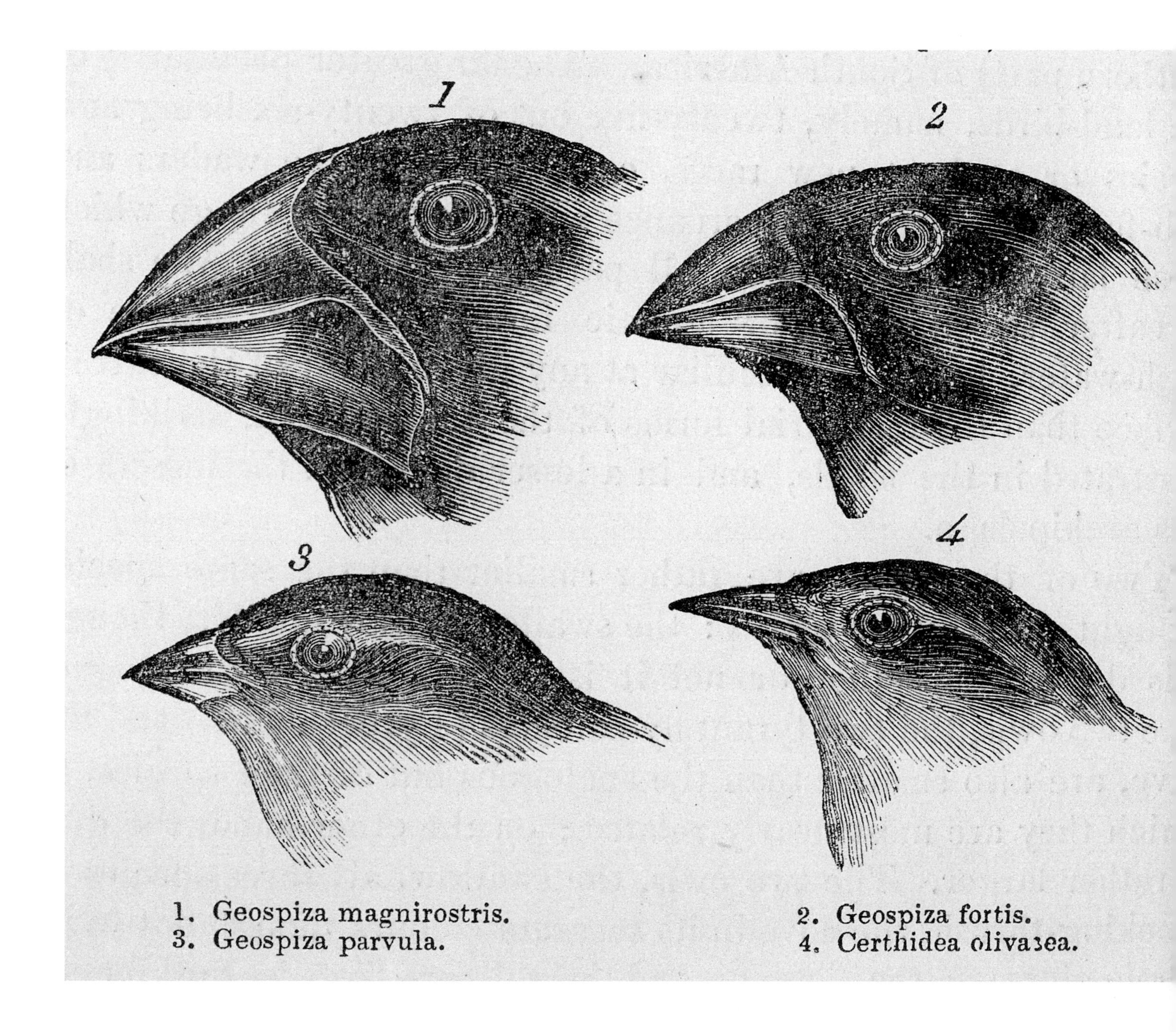

1. Geospiza magnirostris.
2. Geospiza fortis.
3. Geospiza parvula.
4. Certhidea olivacea.

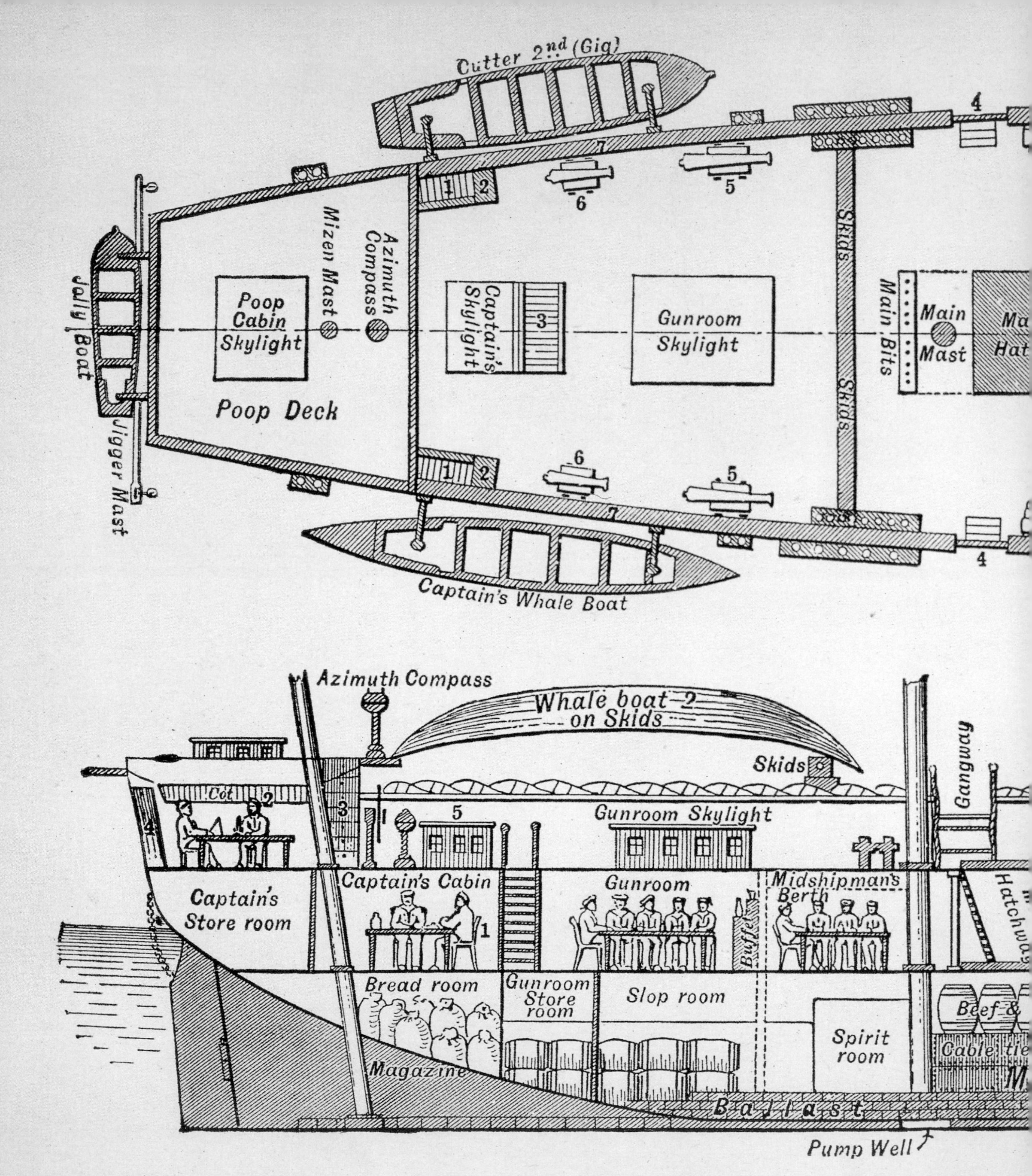

Cutter 2nd (Gig)
4
7
1
2
6
5
Mizen Mast
Azimuth Compass
Skids
Main Bits
Jolly Boat
Poop Cabin Skylight
Captain's Skylight
3
Gunroom Skylight
Main Mast
Poop Deck
Jigger Mast
Captain's Whale Boat
Azimuth Compass
Whale boat 2 on Skids
Skids
Gangway
Cot
Gunroom Skylight
Captain's Store room
Captain's Cabin
Gunroom
Midshipman's Berth
Buffet
Hatchway
Bread room
Gunroom Store room
Slop room
Spirit room
Beef &
Cable tie
Magazine
Ballast
Pump Well

DIAGRAMS OF

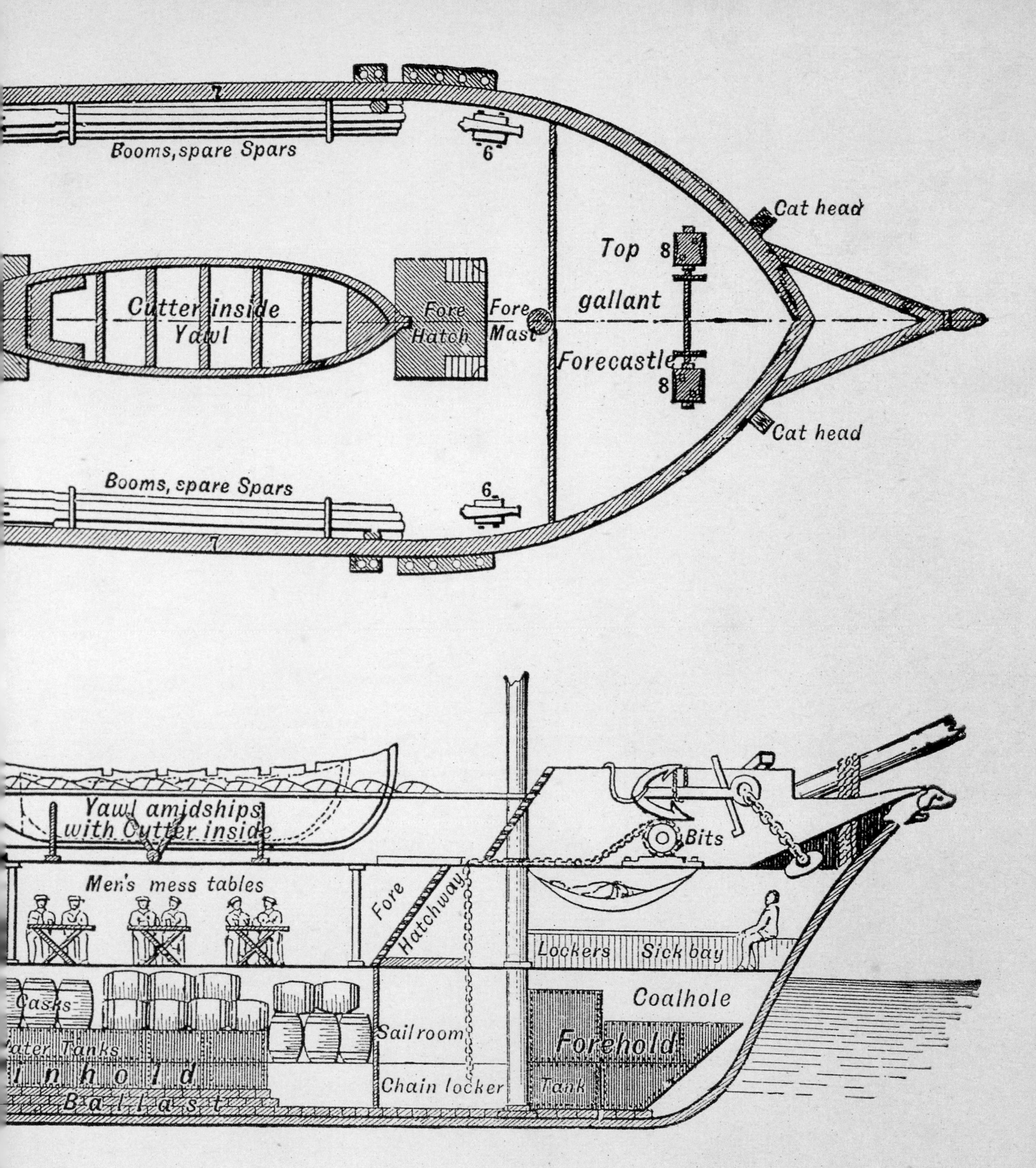
Booms, spare Spars
6
Cat head
Top
8
gallant
Forecastle
8
Cutter inside
Yawl
Fore
Hatch
Fore
Mast
Cat head
Booms, spare Spars
6
Yawl amidships
with Cutter inside
Bits
Men's mess tables
Fore
Hatchway
Lockers
Sick bay
Casks
Coalhole
Sailroom
Forehold
ater Tanks
inhold
Chain locker
Tank
Ballast

IE "BEAGLE."

[To face p. 1.

Phyllostoma Grayi.

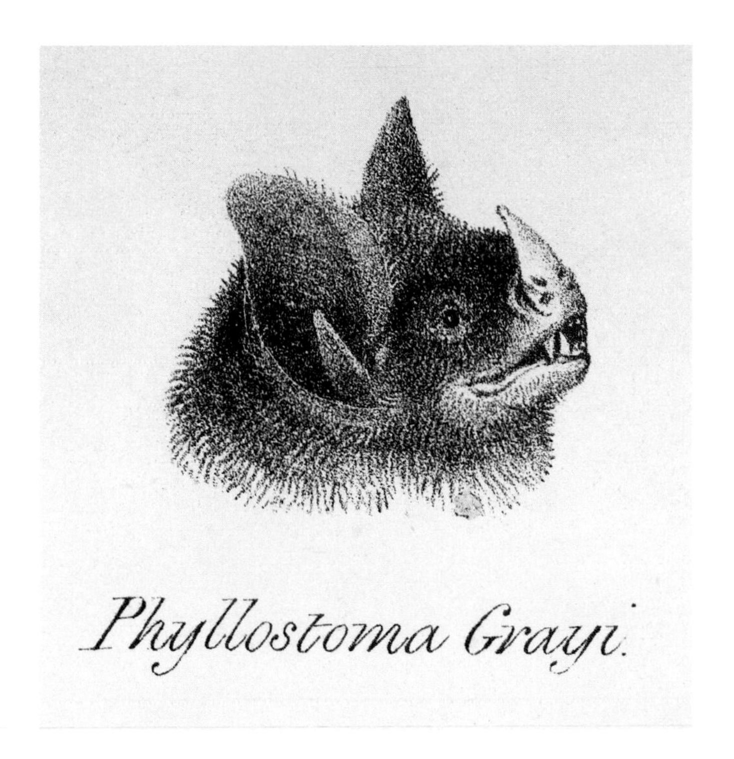

Darwin found this bat *(opposite & above)* – whose given name here is *Phyllostoma grayi* – at Pernambuco, slightly north of Bahia Blanca, in Brazil. 'This species appeared to be common at Pernambuco ...' he wrote. 'Upon entering an old lime-kiln in the middle of the day, I disturbed a considerable number of them: they did not seem to be much incommoded by the light, and their habitation was much less dark than that usually frequented as a sleeping place by these animals.' While the combined length of this particular bat's head and body was only 5 centimetres (2 inches), its wingspan was 25 centimetres (10 inches).

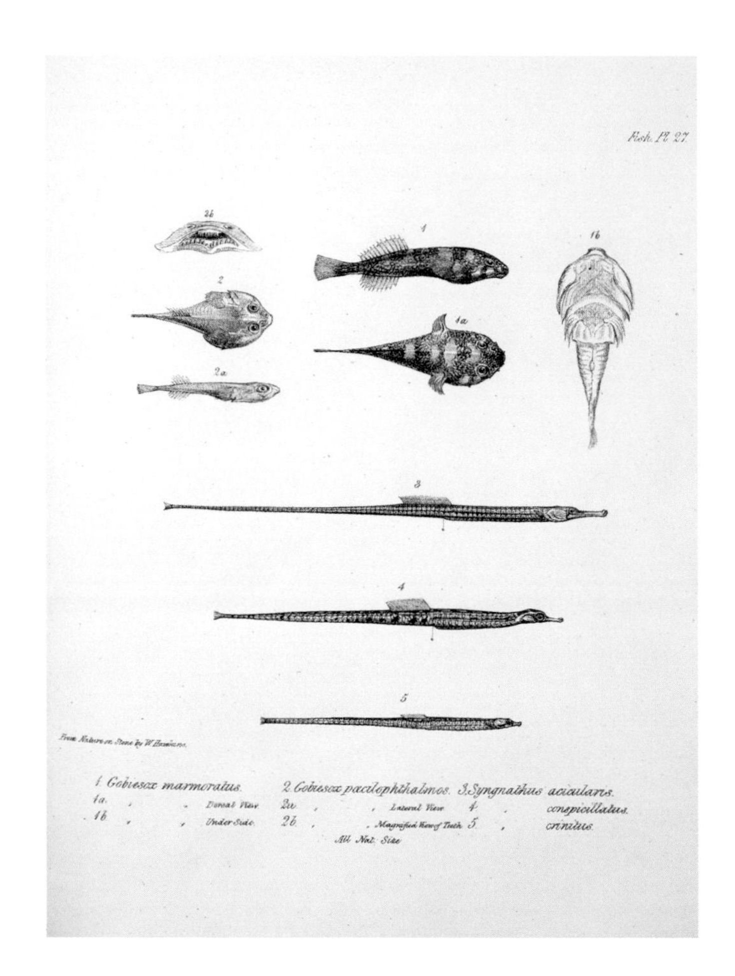

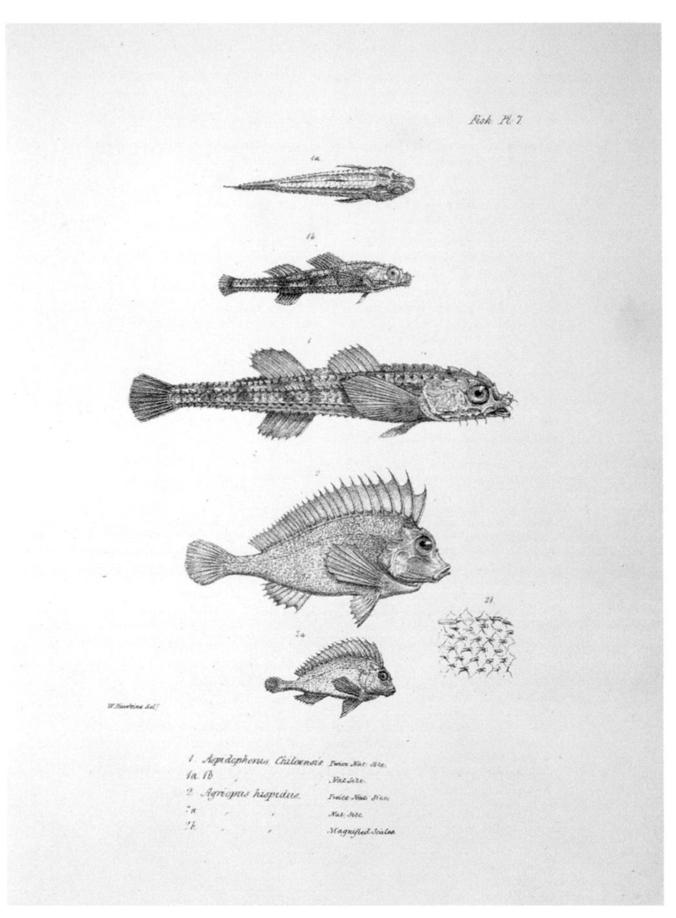

Fish from South American and Tahitian waters *(above left)* including: *Gobiesox marmoratus* (1, 1a, 1b), which Darwin found in the Chiloe archipelago off the west coast of South America; *Gobiesox poecilophthalmus* (2, 2a, 2b), of which Darwin found a single specimen on Chatham Island (Isla San Cristobal) in the Galapagos; *Leptonotus blainvilleanus* (3) from Valparaiso; *Halicaimpus crinitus* (4) from Tahiti; and *Corythoictithys flavofasciatus* (5) from Bahia Blanca. Two tiny species *(above right)* from the Chiloe archipelago are *Agonopsis chiloensis* (1, 1a, 1b) and *Agriopus hispidus* (2, 2a, 2b). Numerous frogs were collected in South America *(opposite)*, including *Pleuroderma bufonina* (1, 1a) from Patagonia, which, Darwin wrote, '... is bred in and inhabits water far too salt to drink'.

Drawn from Nature on Stone by B. Waterhouse Hawkins.

C. Hullmandel Imp.

1. 1 a. *Leiuperus salarius.*
2. 2 a. 2 b. 2 c. *Pyxicephalus Americanus.*
3. 3 a. 3 b. *Alsodes monticola.*
4. 4 a. *Litoria glandulosa.*
5. 5 a. 5 b. *Batrachyla leptopus.*

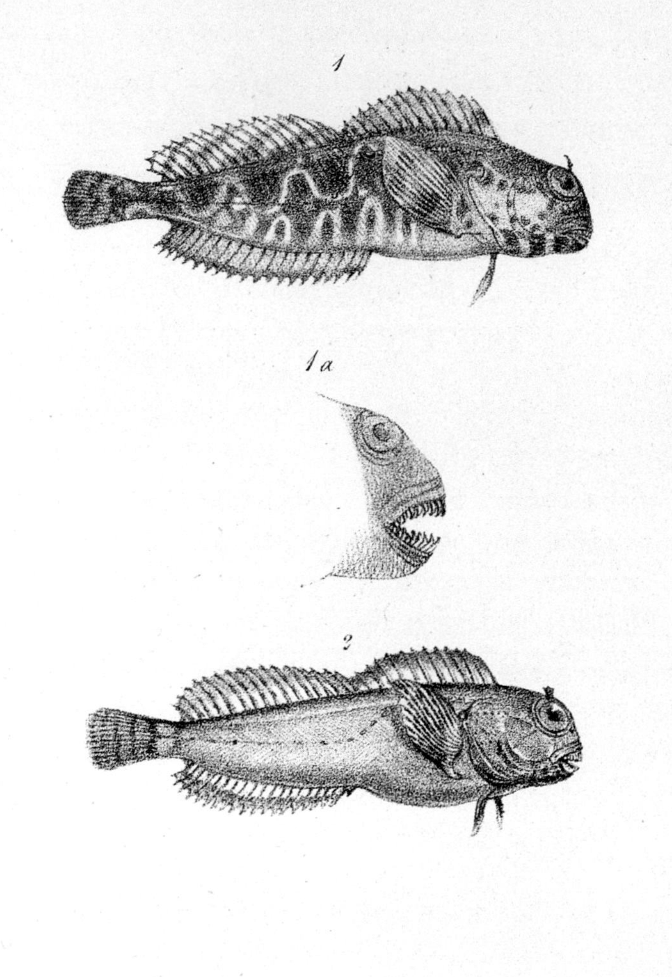

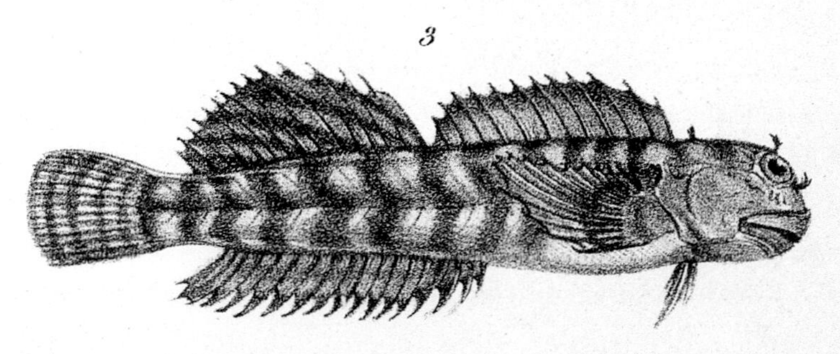

Waterhouse Hawkins del.

1. Blennechis fasciatus. Nat. Size.
1 a. " " Teeth magnified.
2. Blennechis ornatus. Nat. Size.
3. Salarias Vomerinus. Nat. Size.

Three species of fish *(opposite)* including, at the bottom, *Ophioblennius atlantious*, caught early on in the expedition at Porto Praya in the Cape Verde islands. Armed with two long, sharp canine teeth in its lower jaw, it could inflict a nasty wound and, as Darwin noted, did so when it drove its teeth through the fingers of an officer on board the *Beagle*. Among the South American frogs and toads collected *(above)* was *Phryniscus nigricans*, in the centre – now known as *Melanophryniscus stetcheri* – which showed a distinct antipathy to water. At Maldonado, where it was found in the coastal sand dunes, Darwin threw one into a freshwater pool, but it could not swim and had to be rescued. A significant find that extended the range of a genus was *Scorpaena histrio (overleaf)*, a variety of scorpion fish previously only seen on the east coast of America or in the East Indies.

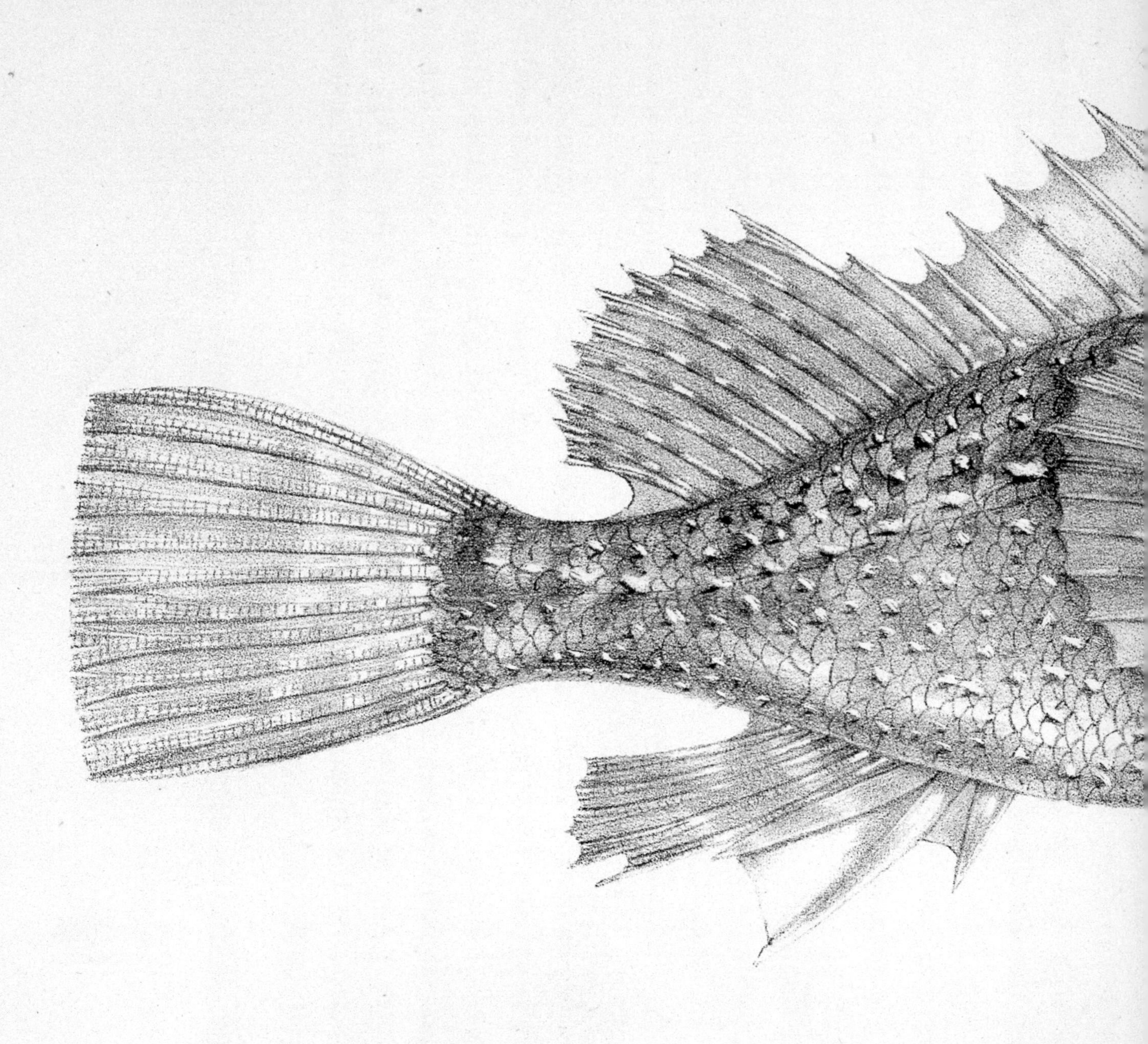

Scorpæna h

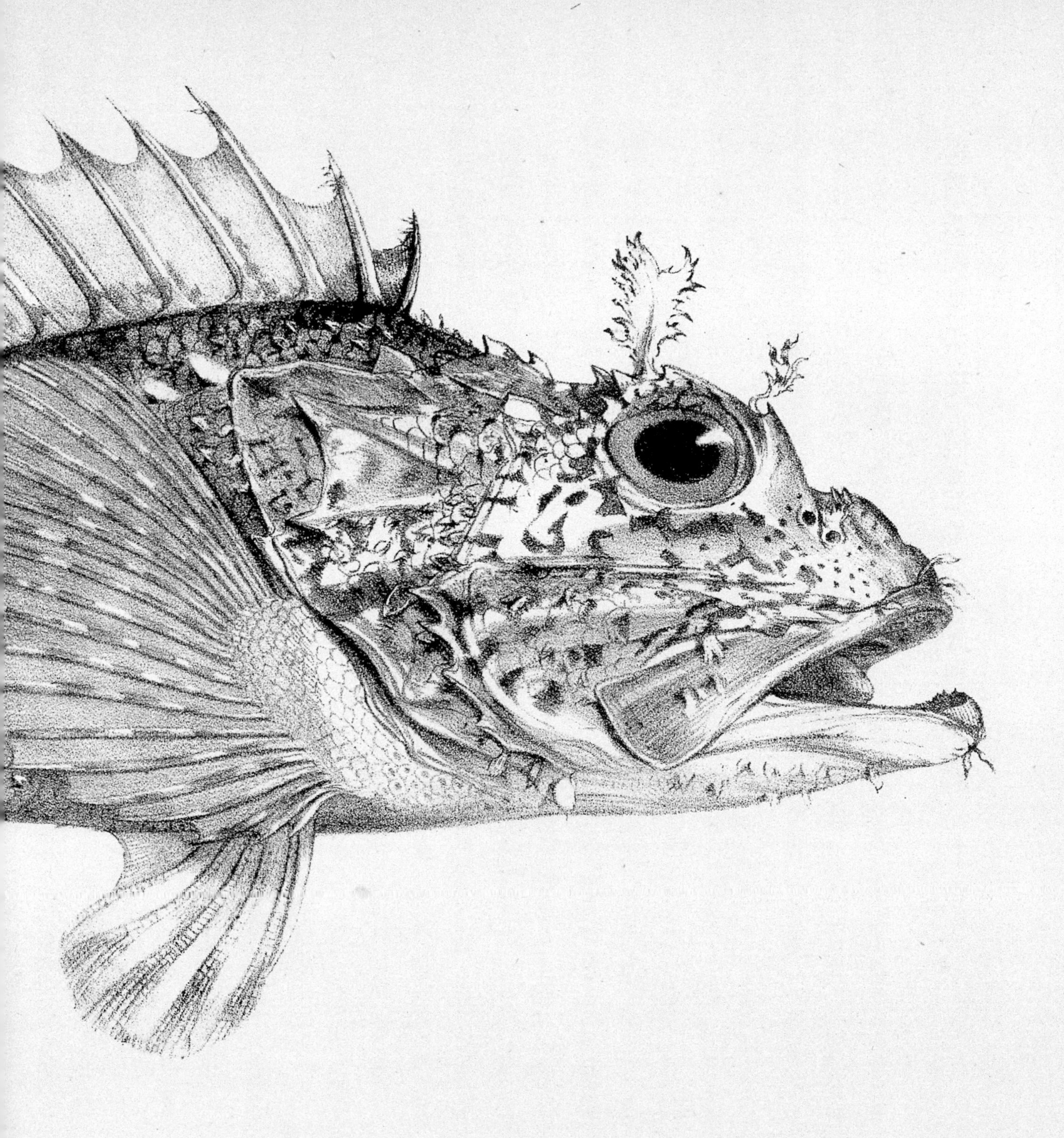

rio. Nat: Size.

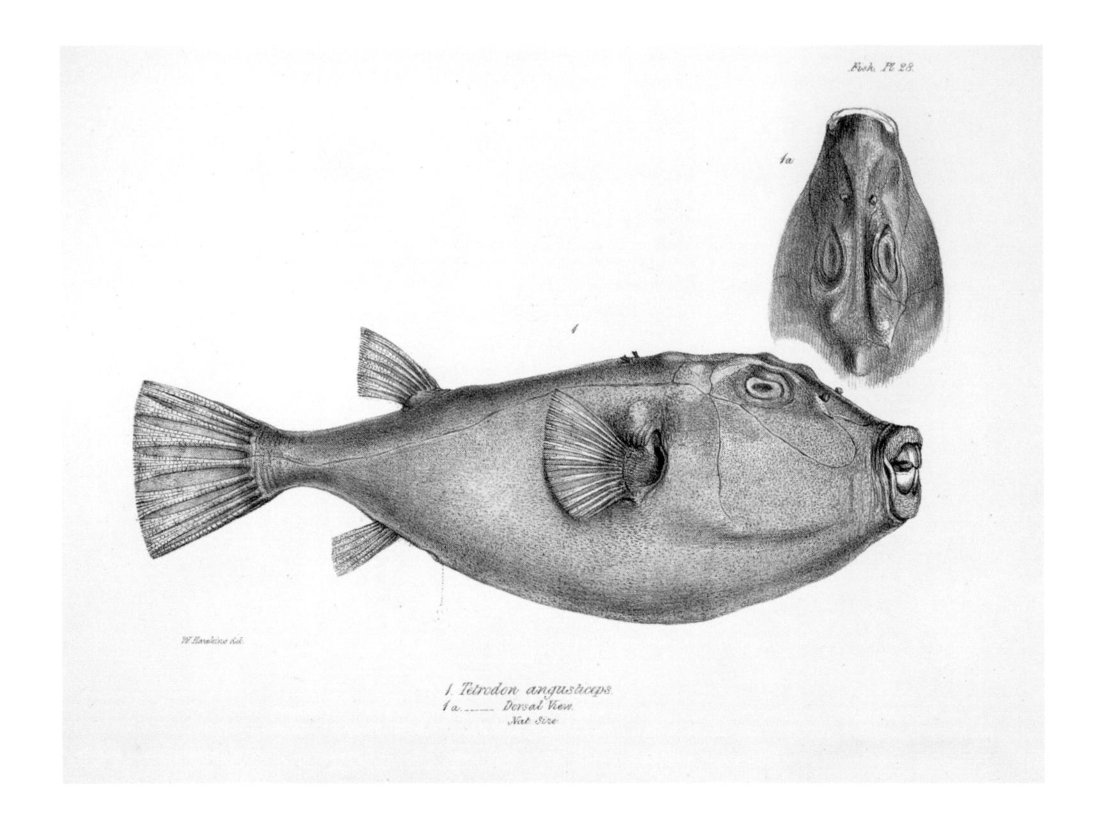
Fish. Pl. 23.
1a
1
W. Hawkins del.
1. Tetrodon angusticeps.
1 a ...... Dorsal View.
Nat. Size

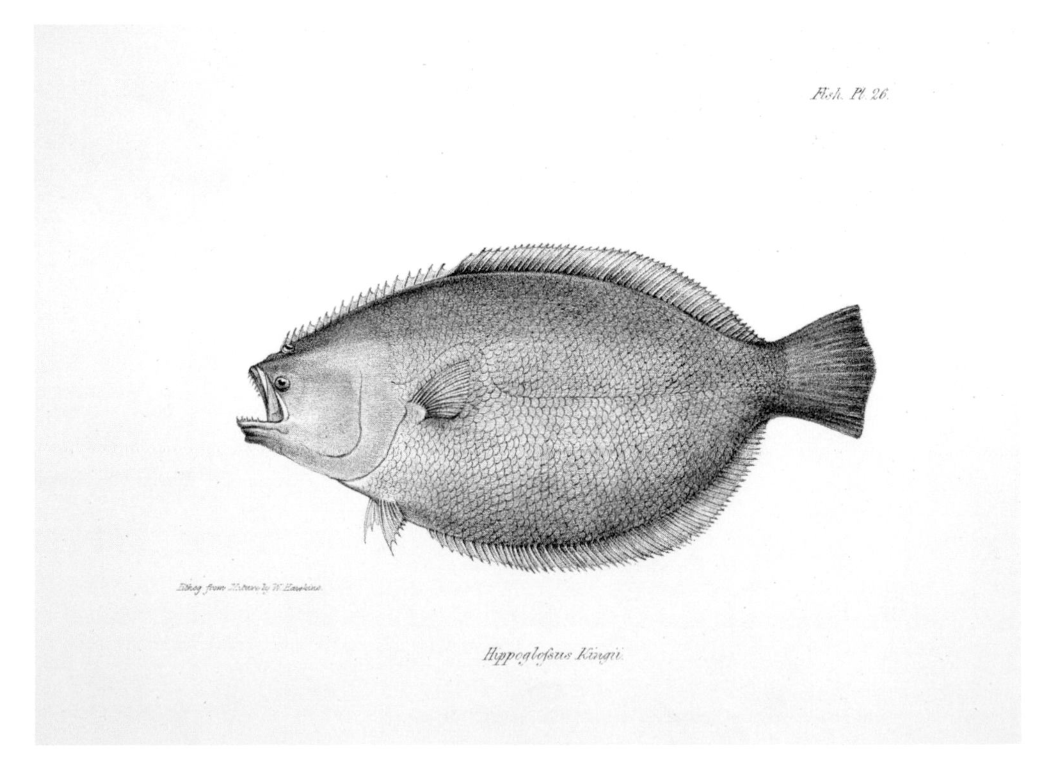
Fish. Pl. 26.
Lithog from Nature by W. Hawkins
Hippoglossus Kingii.

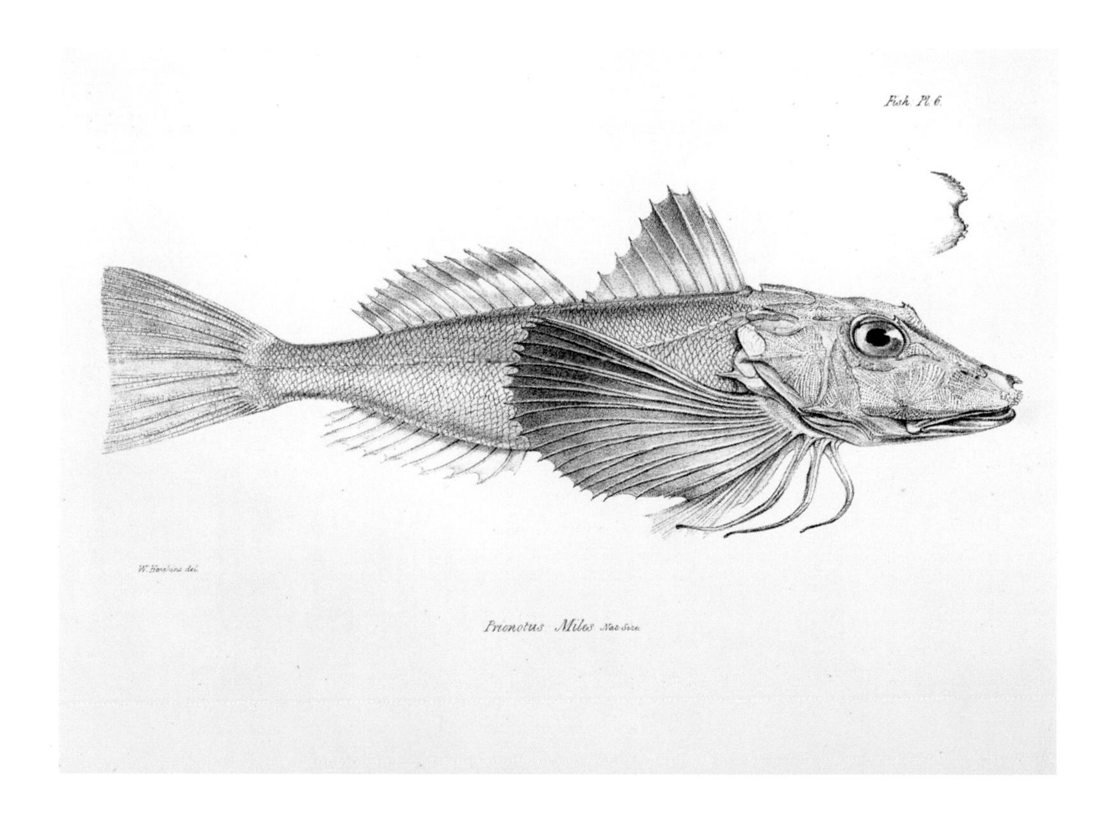

Caught in waters around the Galapagos, this is *Sphoeroides angusticeps (opposite above)*. Darwin also noted that the fish was 'inflatable' – in other words, it was a puffer fish, which when threatened inflates itself by taking in air or water until its body balloons, forcing the spines on the surface to stand erect. Another fish displaying peculiar physical traits is the flatfish *Hippoglossus kingii (opposite below),* with both eyes situated on the left side of its head. It was named after Phillip King, one of the ship's officers, who made a drawing of it for FitzRoy. *Prionotus miles (above)* was a surprise find in the Galapagos, as the genus was thought to inhabit only the Atlantic. Found off the coast of Patagonia, this dolphin *(overleaf)* was named by Darwin *Delphinus FitzRoyi* in honour of FitzRoy, who later returned the honour by naming a mountain in Tierra del Fuego after Darwin.

Delphinus

...tz-Royi.

Two species of lizard from very different parts of the world. *Homonota darwini (above)*, or Darwin's Marked Gecko, came from Port Desire in Patagonia where, Darwin noted, it could be found 'in very great numbers under stones'. He also observed that it made a grating noise when held – it is, in fact, a type of gecko, a common name which mimics the sound the lizard makes. Brown in colour, sometimes tinged with a strong green, it faded in colour after death and, according to Darwin, 'a specimen being kept for some days in a tin box, changed colour into an uniform grey, without the black cloudings ...' The larger lizard *(opposite)* is a Green Tree Gecko, *Naultinus elegans,* originally named *Naultinus grayii*. Of 'fine green' colour, this came from the Bay of Islands in New Zealand, where it lived in trees and was said to make a sound that resembled laughter.

1. *Gymnodactylus Gaudichaudii.*
2. *Naultinus Grayii.*

Craxirex Galapagoensis.

***Birds from Darwin's* Zoology of the Beagle,** with illustrations attributed to John and Elizabeth Gould. From the Galapagos Islands came the hawk, *Craxirex galapagoensis (opposite),* which was tame, like the other birds on the islands including an owl *(above left),* named *Otus galapagoensis.* While in Patagonia, Darwin was keen to find a species of ostrich *(above right),* known in Spanish as 'Avestruz Petise', as earlier naturalists had failed to get a specimen. When one of the crew shot one and it was cooked and brought to table, it was some time before Darwin realised that the bird he had been seeking lay on the plate before him: 'I looked at it, and from most unfortunately forgetting at the moment the whole subject of the Petises, thought it was a two-third grown one of the common sort. The bird was skinned and cooked before my memory returned. But the head, neck, legs, wings, many of the larger feathers, and a large part of the skin, had been preserved. From these a very nearly perfect specimen has been put together, and is now exhibited in the museum of the Zoological Society.' It was named *Rhea darwinii.*

Finches from the Galapagos: *Cactornis assimilis (opposite)* with a beak adapted for feeding on cactus (although the cactus shown here is, incorrectly, a Patagonian species); and *Geospiza fortis (above),* which was found on Charles and Chatham Islands (Isla Santa Maria and Isla San Cristobal). The differing beak shapes of the finches at first led Darwin to classify them into separate subfamilies. Some he called 'Grosbeaks', others were 'Fringilla', or 'true finches', while those which fed on cactus belonged to the 'Icterus' family which included blackbirds and orioles. Back in London, the ornithologist and publisher John Gould classified all of Darwin's bird specimens, including the Galapagos finches, which he said were a completely new group, each bird representing a separate and mutually exclusive species of finch.

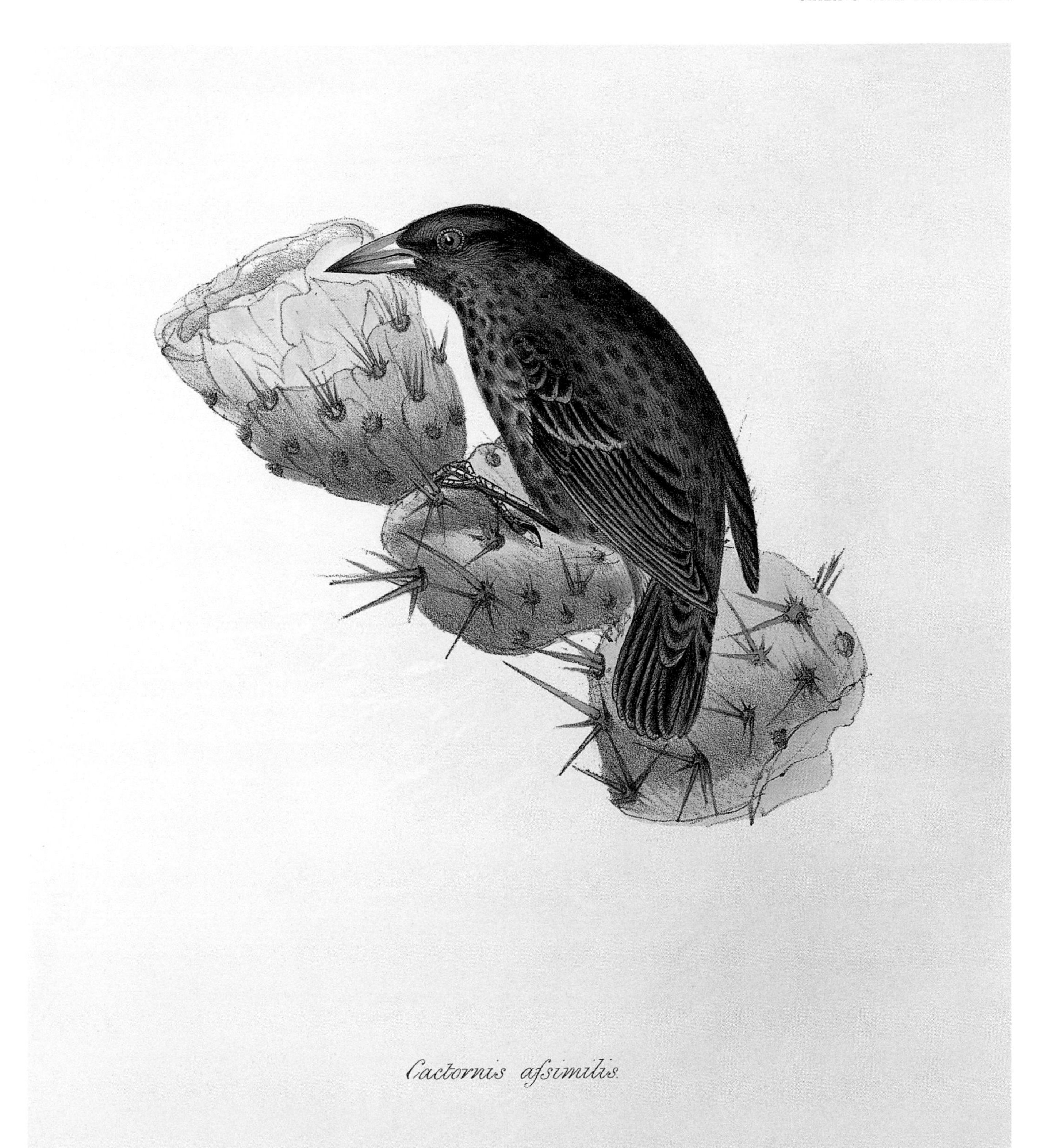

Cactornis aſsimilis.

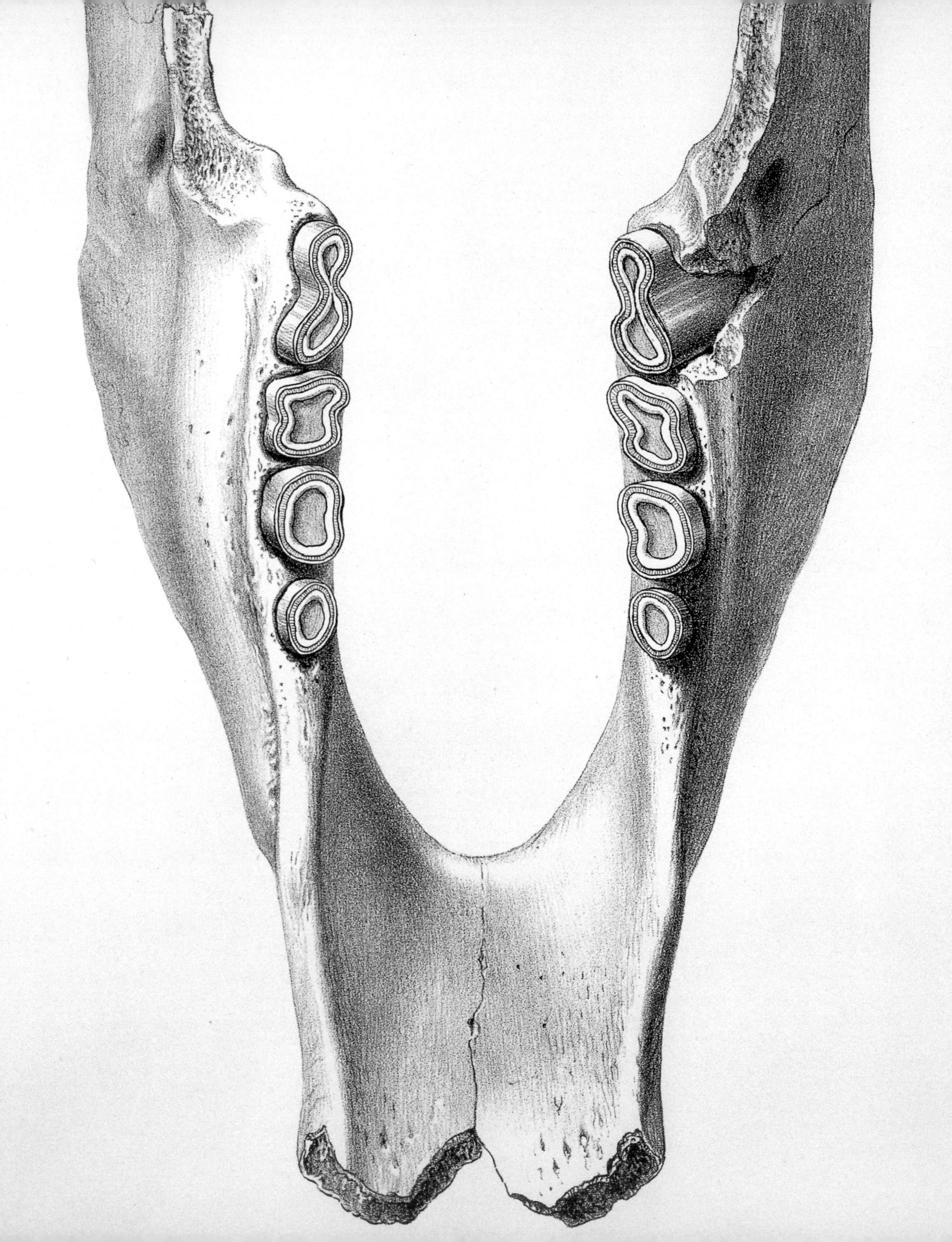

In September 1832, when the *Beagle* was anchored at Bahia Blanca, Darwin noticed several fossilised animal bones embedded in soft rock at Punta Alta. He began to unearth them, sometimes working through the night, yielding the incomplete skeletons of three large animals, including the jaw bone of *Mylodon darwinii (opposite)*. These discoveries puzzled Darwin: they were clearly the bones of extinct animals, yet they resembled those of smaller, living creatures. This relationship between past and present species was another link in the chain that would lead him to his evolutionary theory. At Port St Julian in Patagonia, Darwin found the fossilised remains of another mammal, a llama-like creature about the size of a camel, *Macrauchenia patachonica*. The larger bones shown here *(below)* are its cervical vertebrae.

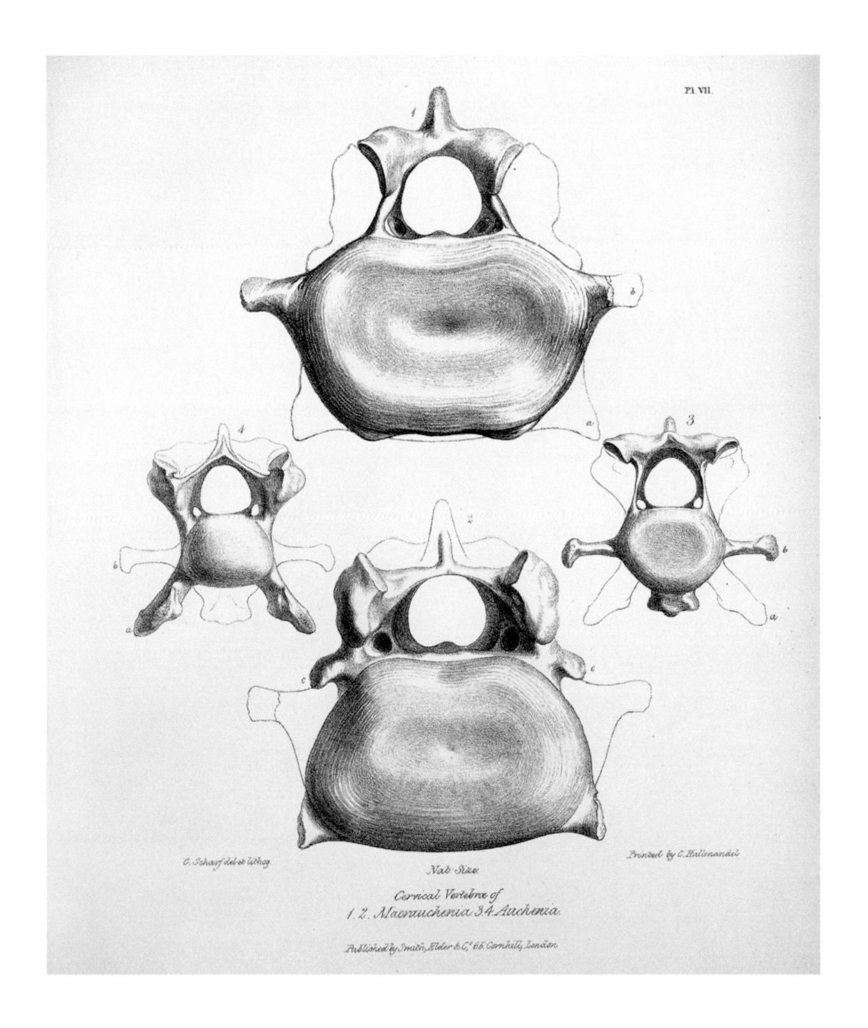

Amazonia and Beyond 1848–1862

Alfred Russel Wallace & Henry Walter Bates

Included in the two volumes of notes by Henry Walter Bates on Amazonian insects are hundreds of his exquisite drawings *(opposite)*, made during jungle expeditions and at several bases along the Amazon.

IN THE SPRING OF 1858, A 34-YEAR-OLD WELSHMAN LAY IN A COT in a palm-thatched house on the tiny island of Ternate in the Moluccas, now part of Indonesia, alternately shivering and sweating through an attack of malarial fever. As he drifted in and out of semi-consciousness, Alfred Russel Wallace mulled over in his mind the vast diversity of animal life that he had encountered over the previous 10 years, here in the islands between the Indian and Pacific Oceans and earlier in the Amazonian rainforests. How on earth had all these species, so beautifully adapted to their environment, come into being? Eventually, between the worst bouts of fever, he scribbled down his ideas in a 4,000-word essay and sent it off to his associate Charles Darwin in England together with a covering letter asking Darwin if he thought it was worth submitting for publication in a learned journal.

When Darwin received it three months later he was dumbfounded. Here, in this short manuscript, Wallace had summarised all the main principles of Darwin's own ideas on evolution, the theory of natural selection which he had been working on for more than 20 years since the *Beagle* voyage. If Wallace's manuscript was published alone, Darwin would be pipped to the post. A solution was arrived at by Darwin's friends the geologist Sir Charles Lyell and the botanist Joseph Hooker – on 1 July 1858, Wallace's paper was presented, together with a summary of Darwin's own findings, to the Linnean Society in London. Neither of the authors were present, Wallace because he was in New Guinea and Darwin because one of his children had just died of scarlet fever. The solution, on the face of it, seemed even-handed to both parties, but immediately afterwards Darwin began feverishly to prepare his ideas for publication. The result was his book, *On the Origin of Species*, which appeared in 1859 and ran through six editions over the next 13 years. Although Wallace's pioneering work was acknowledged, at least in the earlier editions, credit for the theory became concentrated increasingly on Darwin and less on Wallace, until eventually Wallace's contribution became almost forgotten. Amazingly, Wallace felt no resentment and, in his later life, seemed content to be remembered for his accounts of his remarkable travels, his skill as an animal collector and the founder of what we now call biogeography, the study of animal distribution, rather than as one of the originators of what he called 'Darwinism'.

No 1
No 2
No 3
No 4
No 5
No 6
No 7
No 8
No 9
No 10
No 11
No 12
No 13
♀
No 14
No 15
No 16
No 17
No 18
♀
No 19
No 20
No 21
No 22
No 23
No 24

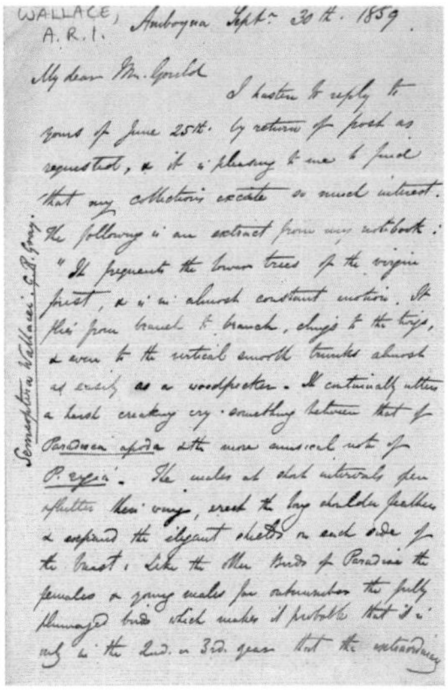
WALLACE, A.R.I. Amboyna Septr. 30th. 1859.

My dear Mr. Gould

I hasten to reply to yours of June 25th. by return of post as requested, & it is pleasing to me to find that my collections excite so much interest. The following is an extract from my note-book: "It frequents the lower trees of the virgin forest, & is in almost constant motion. It flies from branch to branch, clings to the twigs, & even to the vertical smooth trunks almost as easily as a woodpecker. It continually utters a harsh creaking cry, something between that of Paradisea apoda & the more musical note of P. regia. The males at short intervals open & flutter their wings, erect the long shoulder feathers & expand the elegant shields on each side of the breast. Like the other Birds of Paradise the females & young males far outnumber the fully plumaged birds which makes it probable that it is only in the 2nd. or 3rd. year that the extraordinary

Semioptera Wallacei: G.R. Gray.

***A letter to the ornithologist John Gould** (left)* from Alfred Russel Wallace, written in Amboyna in 1859. The two men shared a fascination with Birds of Paradise. John Gould's first volume of *Birds of New Guinea*, commenced in 1875, included several plates based on Wallace's specimens. They are some of Gould's most extravagant and colourful illustrations and are a fitting tribute to one of the world's greatest animal collectors.

Wallace was born at Usk in south Wales in 1823, the seventh child of a bookish but unsuccessful businessman who moved his family to Hertford in 1828. Alfred left school at 14 and worked with two of his brothers, William, a surveyor, and John, a builder, until 1844 when he got a job teaching at the Collegiate School in Leicester. He was already interested in botany, but at Leicester he met a local lad, Henry Walter Bates, who was also fascinated with natural history, but in his case particularly entomology. The two young men fired one another's enthusiasm and decided to devote their lives to their hobby – and if possible make a living from it.

As they were not independently wealthy (unlike Darwin), the only way Wallace and Bates could make money in the field was as collectors to supply the seemingly insatiable mid-19th century appetite for private and public collections of new and exotic species from previously unexplored parts of the world. They parted company for a couple of years when Wallace went back to Wales to sort out his brother's surveying business after William's death in 1846. The rapidly extending railway system made surveying temporarily very profitable and Wallace soon accumulated enough funds to finance the grand plan he and Bates had hatched to collect in South America around the Amazon.

With assurance from the authorities at the British Museum that any specimens they collected would find a ready market, Wallace and Bates sailed from Liverpool in April 1848 to Pará (present day Belém), near the mouth of the Amazon. Apart from a journey of about 160 kilometres (100 miles) down the Rio Tocantins they spent the first year in Pará learning the local customs and what was to become their lifelong trade as animal collectors. The diversity of the animals, and particularly insects, in the forests and around the rivers was amazing. In their first two months of collecting Wallace could report to Samuel Stevens, their agent in London, that they had collected no less than '... 553 species of Lepidoptera ... 450 beetles, and 400 of other orders ...'

Following their joint expedition along the Tocantins, Wallace and Bates generally journeyed separately, presumably so that they could cover a bigger area. Wallace stayed in South America for four years, returning to England in 1852 partly as a result of the death of his younger brother Herbert who had joined him in 1849. Herbert had accompanied

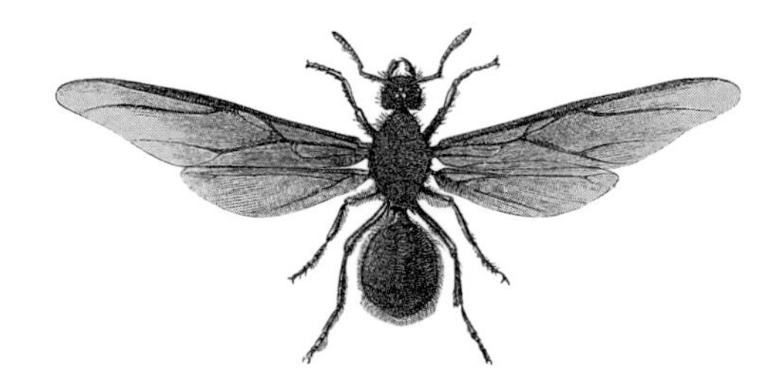

Alfred on his first major expedition up the Amazon as far as Manaus, but had died of yellow fever in Pará in June 1851. Alfred had continued his journeys in the upper Amazonian region, concentrating particularly on the Rio Negro and the Rio Vaupés. He assiduously collected plants and animals, particularly birds and insects, some of which were despatched back to Stevens to be sold to fund further forays. He collected fish from the rivers using a poison called 'timbo' obtained from the roots of *Paullinia pinnata* and *Lonchocarpus nicou*. He described and drew some hundreds of species of fish and accumulated an impressive collection. Tragically, the specimens were destroyed on his way home when the ship he was in caught fire. Luckily, Wallace managed to save a tin box from his burning cabin, which held his notes and drawings of the fishes, as well as his notes on the survey of the Negro and Vaupés, including sketches of palms and places he had visited. The shipwrecked survivors spent 10 days adrift in open boats before being picked up and brought back to England.

When he landed at Deal in Kent, Wallace had, in his own words, '£5 and a thin calico suit'. His expenses in South America had been covered by the sale of the few specimens he had sent back to Stevens, most of which eventually found their way to The Natural History Museum in London, but he had hoped to make another £500 from the sale of the specimens lost in the fire. Fortunately, Stevens had insured them for £150, which was some compensation. More seriously, Wallace had hoped to use his notes and specimens for a major publication which would surely make his name. This was now impossible, and when his published account of his Amazonian experiences was less than a resounding financial success, Wallace's next – and already planned – collecting excursion became doubly essential.

As his new target Wallace chose the Malay Archipelago. Meanwhile, his friend Bates stayed in South America for a total of 11 years making extensive and sometimes dangerous collecting expeditions from a number of bases including Santarém on the lower Amazon and Ega and São Paulo de Olivença on the upper Amazon and the Solimões Rivers. By the time he returned to England in 1859, still only 34 years old, Bates had collected, by his own estimate, some 712 species of mammals, reptiles, birds, fishes and molluscs, and about 14,000 species of insects, of which no less than 8,000 were previously unknown.

***The notebooks of Wallace** (opposite)* contain detailed notes and illustrations on the insects and birds he observed in many different locations around the Malay Archipelago.

Unlike Wallace's Amazon book, Bates' superbly written account of his time in South America, *The Naturalist on the River Amazons,* published in 1863, was immediately recognised as one of the classics of scientific narrative and exploration, with wonderful descriptions of the topography, climate and the nature and customs of the native peoples as well as of the flora and fauna. Bates had kept meticulous notes, and extracts of many of the letters he had sent back to scientific colleagues in England were published in learned journals. He also wrote several important papers during the 27 years he served as Assistant Secretary of the Royal Geographical Society.

However, his main claim to zoological fame is his recognition of a phenomenon, to this day called 'Batesian mimicry', which he, and many other 'Darwinians', felt provided support for the idea of natural selection. This was the very close resemblance in colour patterns and even in superficial morphology between butterfly species that are palatable to bird predators and other butterfly species that birds find extremely unpleasant or even harmful to eat. By mimicking the noxious species, the harmless ones gain protection from predation despite their palatability – a superb example, thought the evolutionists, of the powers of natural selection.

Wallace, in the meantime, was making his own unique contribution to zoology. The mass of islands between Malaya and Australia had been visited by Europeans many times since the 16th century, but apart from Java almost none of them had been thoroughly investigated by naturalists. There was every likelihood that they harboured numerous new and interesting species that would enable Wallace to compensate for his fairly disastrous Amazonian venture. With hopes high, he left England in March 1854 and arrived in Singapore on 20 April. He spent six months on the Malay Peninsula collecting around Singapore and Malacca (now called Malaka) and then set out on his eight-year exploration of the islands, during which he was to cover some 22,400 kilometres (14,000 miles), moving his base more than 90 times. He visited almost every group of islands in the area, his most easterly voyages taking him to the Aru Islands and New Guinea. As before, Wallace periodically sent his collections back to Stevens, the first consignment of about 1,000 insects from Malacca containing at least 40 new species. But by the time he came back to England, arriving

in London on 1 April 1862, these numbers had paled into insignificance. He had collected in total a prodigious 125,000 specimens, most of them beetles but including many other invertebrates, birds, mammals, amphibians, reptiles and fishes. It is still recognised as one of the most important collections ever made and has formed the basis of numerous scientific publications, including many by Wallace himself.

Even before he published his account of his travels, *The Malay Archipelago*, in 1869, he had published 18 papers on the material, while almost 2,000 new beetle species and hundreds of new butterflies had been described from his collections by other naturalists. In contrast to his account of his Amazonian experiences, *The Malay Archipelago* was very well received. This was partly because of the acknowledged importance of the collections, and Wallace's association with Darwin's theory, but the account was also well written, and there was an inherent interest in the region he was describing.

In purely scientific terms, and apart from his thoughts on natural selection, the main significance of Wallace's work resulted undoubtedly from his biogeographic observations. He came to the conclusion, still accepted today, that the archipelago represented the frontier between two great faunal provinces, an Indo-Malayan one to the west and an Australian one to the east. The idea behind this, that the distribution of animal species, and indeed whole faunas, might be related to both geological and what we would now call evolutionary history, was quite novel at the time. Wallace expanded this idea in 1876 when he published *The Geographical Distribution of Animals*, a book which established him as the father of modern biogeography. In the specific case of the Malay Archipelago, Wallace believed that the two zones could be separated by an imaginary line running between the Philippines, Borneo and Java which belonged to the Indo-Malayan area, and Celebes, the Moluccas, Timor and New Guinea, with faunas belonging to the Australian region. Although this 'line' has been moved backwards and forwards by biogeographers in the ensuing century and a half, it is still called the 'Wallace Line' in deference to Wallace's original recognition of it.

For the average reader, however, the main attraction of *The Malay Archipelago* was the chapters devoted to the subjects of the book's subtitle, 'the Land of the Orang-utan and the Bird of Paradise ...', both the focus of great

Miss F. Chant, Parkstone, Photographer
Emery Walker Ph.sc.
Alfred R. Wallace

On his return to England in 1862, Wallace *(opposite)* lived contentedly with his wife, largely on the income resulting from the sale of his Malay specimens and from writing scientific articles and several books. His old age was spent mainly devoted to gardening and he died at Broadstone, Dorset in 1913 at the age of 90.

fascination. Wallace had spent almost three months in Sarawak primarily to observe and collect orang-utans, which would fetch a very high price in Britain. He collected most of his Bird of Paradise specimens from the Aru Islands but did not find them easy to track down. 'It seems,' he wrote, 'as if Nature had taken precautions that her choicest treasures should not be made too common and thus undervalued.' So despite his best efforts he managed to obtain only six of the 18 species known at that time and one new one, Wallace's Standard Wing, which G. R. Gray of the British Museum named *Semioptera Wallacei* in his honour.

Wallace even invested the princely sum of £100 in two living Lesser Birds of Paradise, *Paradisea papuana*, which he successfully brought back to England, one of them surviving in London Zoo for almost two years. But Wallace's Birds of Paradise were brought to the notice of a much wider audience when they were used by the ornithologist and publisher, John Gould, in his last great illustrated work, the *Birds of New Guinea.*

Once back in England – having lost the post of Assistant Secretary of the Royal Geographical Society to his friend Bates, and later failing to obtain the directorship of the newly established Bethnal Green Museum – Wallace made a living from investments, but particularly from lecturing, including tours in America, and writing. He wrote a number of very successful books, including his autobiography, *My Life,* which appeared in 1905. His interests were extremely varied, ranging from mesmerism and spiritualism, which led him to the conclusion that man was excluded from the process of natural selection, to the possibilities of life on other planets.

Despite his lack of formal qualifications, Wallace was offered honorary degrees by several universities and medals by scientific societies. He mainly refused the degrees and accepted the medals, though when he was offered the prestigious Royal Medal of the Linnean Society he jokingly told a friend: 'A dreadful thing has happened! Just as I have had my medal-case made, regardless of expense, they are going to give me another medal!' And in 1892, against his wishes, he was elected a Fellow of the Royal Society, the ultimate accolade for a scientist. His contributions to science may be less well-remembered than those of Charles Darwin, but Wallace is certainly remembered as a far more impressive collector.

Ever since they were discovered by Europeans in the 16th century, Birds of Paradise have been widely admired for their startling coloration and the remarkable plumage development in the males, the females usually being relatively plain. During his time in the Malay Archipelago, Wallace was understandably keen to observe and obtain as many specimens as possible. Among those he observed were the Greater/Arfak Six-plumed Bird of Paradise *(above)*, called *Parotia sexpennis* by Wallace, and the Paradise Pie or *Astrapia nigra (opposite)*. These were painted by Gould for his *Birds of New Guinea*.

ASTRAPIA NIGRA.

J.Gould & W.Hart del. et lith.
Walter Imp.

The Papuan Bird of Paradise *(opposite), Paradisea papuana* to Wallace but now *Paradisaea minor*. Wallace found this magnificent species on the mainland of New Guinea as well as on several of the nearby islands. He not only obtained well-preserved dead specimens, but also brought the first living ones to England, despite the problem of finding enough cockroaches to feed them. Wallace resorted to setting traps on the ship and stocked up with tins full of biscuits during a stop in Malta . Eventually '... they arrived in London in perfect health, and lived in the Zoological Gardens for one and two years respectively, often displaying their beautiful plumes to the admiration of the spectators.' He did not find such good specimens of the Bare-headed Bird of Paradise *(above left)* or the Magnificent Bird of Paradise *(above right)*.

109
½ nat. size

As Wallace travelled the Rio Negro and the Rio Vaupés, he sketched all the fish he could. These drawings survived the fire that destroyed the rest of his collections and are all that remain of his detailed recording of the animal life of the Amazon basin. *Pterophyllum scalare (opposite)* – called the butterfly fish by Wallace, but now more commonly know as the angelfish – was first introduced to Europe for use in aquaria in the early 1900s. It was an immediate hit, and now hundreds of varieties are commercially available. The black-lined leporinus, *Chalceus nigrotaeniatus (above),* is also available in the aquarium trade. Wallace sketched this fish on the Rio Negro, where it was known to the locals as 'uaracu murutinga'.

Catfishes are incredibly diverse, and especially so in the tropics. With a huge mouth full of very tiny teeth, *Asterophysus batrachus (below)* is one of the peculiar-looking driftwood catfishes. Wallace found it was called 'mamyacú' along the upper Rio Negro. Other catfishes sketched by Wallace look more menacing *(opposite)*. This still unidentified member of the family Doradidae was called 'caracadú' along the upper Rio Negro. Wallace sketched many members of this family, called the talking catfishes because they make an audible sound when removed from the water.

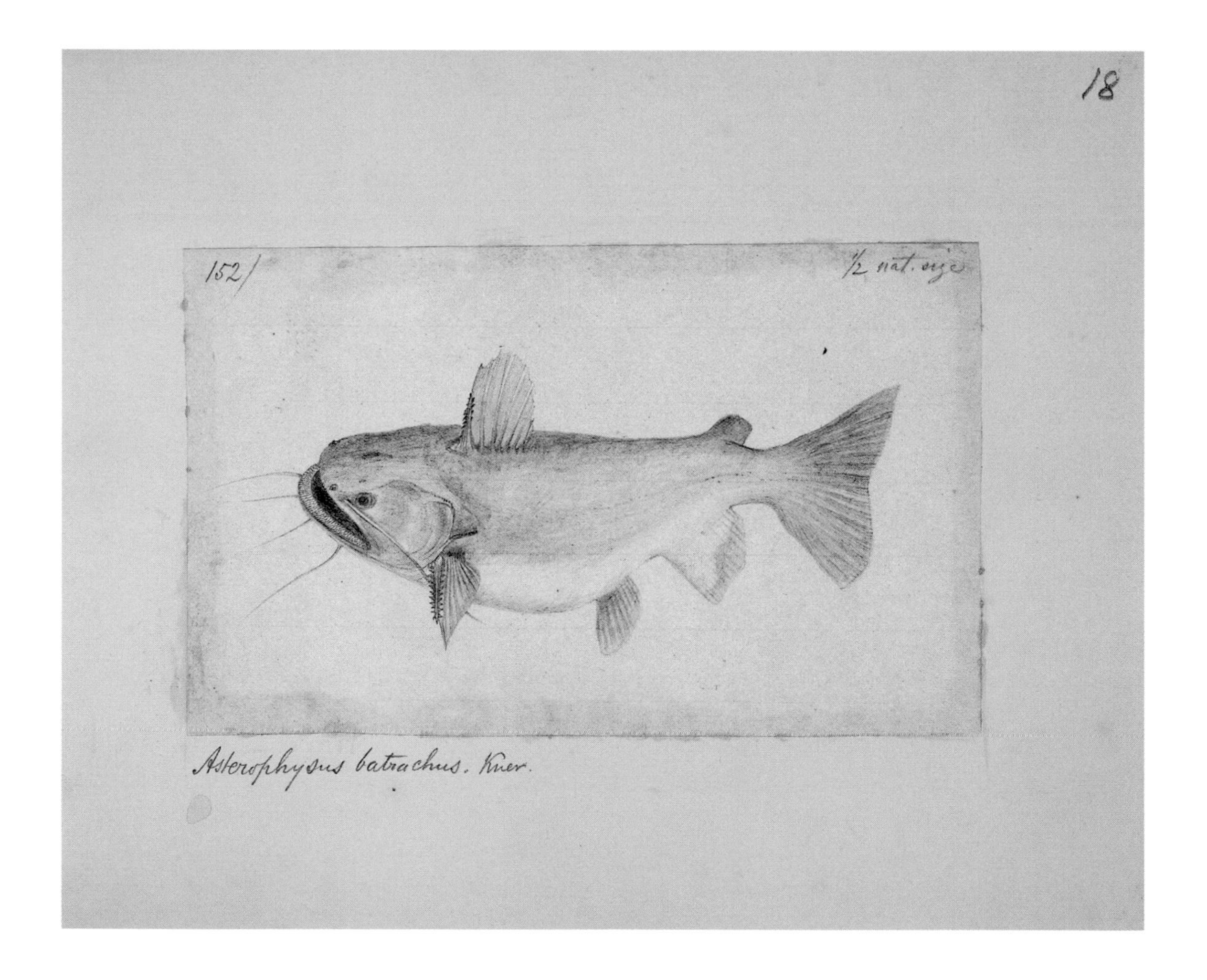

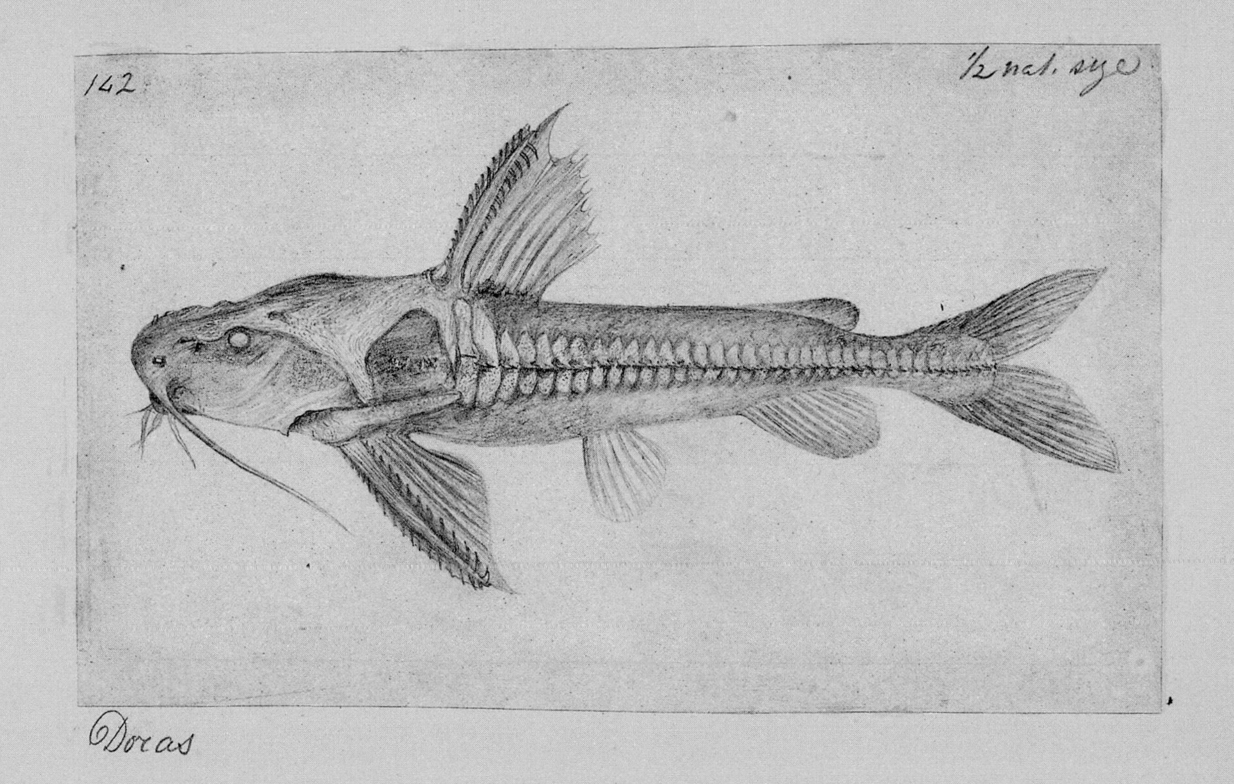
142
½ nat. size
Doras

26

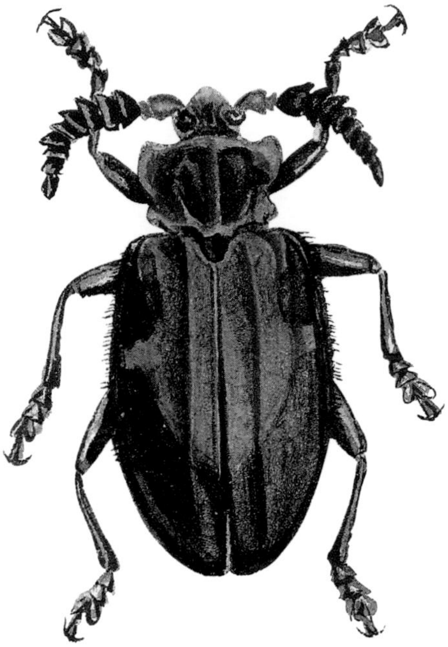

Bates was very interested in Coleoptera, or beetles, the largest insect order, containing well over 250,000 species. Here *(right)* he illustrates the extremes of beetle evolution: on the bottom and middle rows, five relatively primitive predatory tiger beetles (family Cicindelidae); at the top, two plant-eating cerambycid or 'longhorn' beetles, so called because of their unusually large antennae *(above)*.

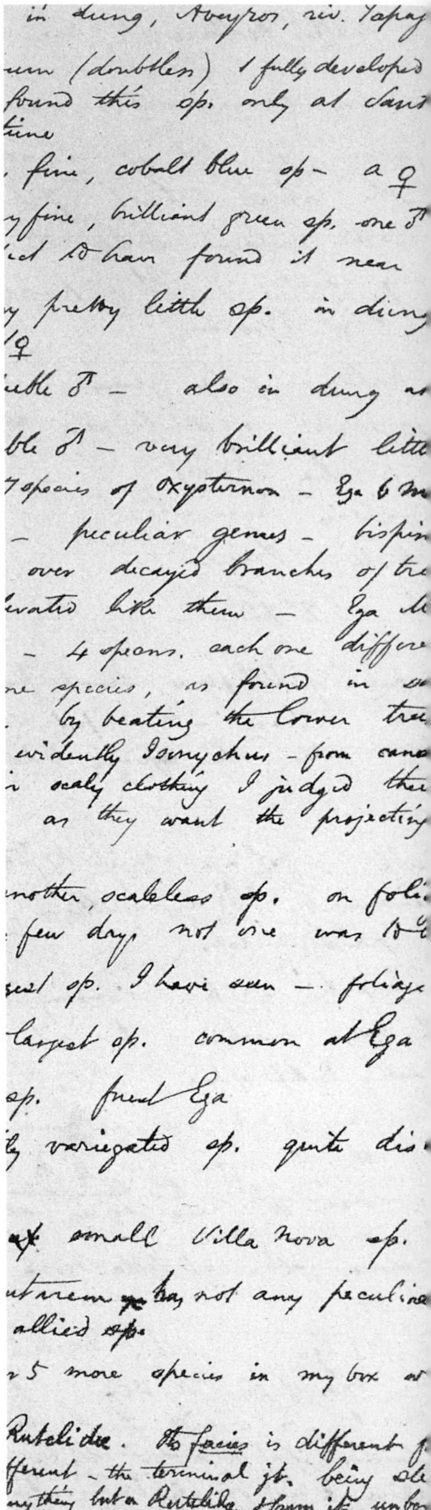
in dung, Aveyros, riv. Tapaj
um (doubtless) 1 fully developed
found this sp. only at Sant
time
fine, cobalt blue sp. – a ♀
y fine, brilliant green sp. – one ♂
ct to have found it near
y pretty little sp. in dung
/♀
uble ♂ – also in dung a
ble ♂ – very brilliant littl
7 species of Oxypternon – Ega 6 m
– peculiar genus – hispin
over decayed branches of tre
levated like them – Ega M
– 4 speans. each one differe
ne species, as found in
by beating the lower tre
evidently Isonychus – from cano
scaly clothing I judged the
as they want the projecting
nother, scaleless sp. on foli
few days not one was to
sest sp. I have seen – foliag
largest sp. common at Ega
sp. found Ega
ly variegated sp. quite dis
small Villa Nova sp.
utarem – has not any peculiar
allied sp.
5 more species in my box
Rutelidae. Its facies is different
fferent – the terminal jt. being sle
nything but a Rutelida, & from its unbo

ly 1852 [illegible] any one
fully desc. ♂ & ♀ – Santarem
where it flies into houses in the
Ega, flying in the forest
– not very rare at Ega – also
ha
Ega, rare
, rare
in clearing. Ega, rare
56
metatho. & cordate abd.
same manner as Cryptoceri & with
156
but I think likely to be all vars
part of forest at Ega. I take
ated mentum, claws, tarsi &c.
t first to be Cerasepis, but they
of mid. of hind margin of
once abundant in forest at Ega
een.
s Ega
from its neighbours – not rare
ly distinct from its neighbours
distinguish it at once from
I will tackle after I have studied
hat of any Rutelidae I have, & its
r &c. In its labrum, mandibles &c.
hind edge of proth. it must come

[illegible] fallen fruits of the Wixi (a hardish pulp) or Muru
palm. – Ega – both varieties I think a
[illegible] – Doubtless ♂ & ♀ as the more brillian
are all
genus Ch
pre with
cleft ext. for
extern.
the feet
feet.
under
an ove
too de
Paris E
but I h
those of
Wixi –
the same sp.
together w
Ega
being gre
tarsi, cleft
my species
being m
prolonged
– & the
into the
abunda
still mo

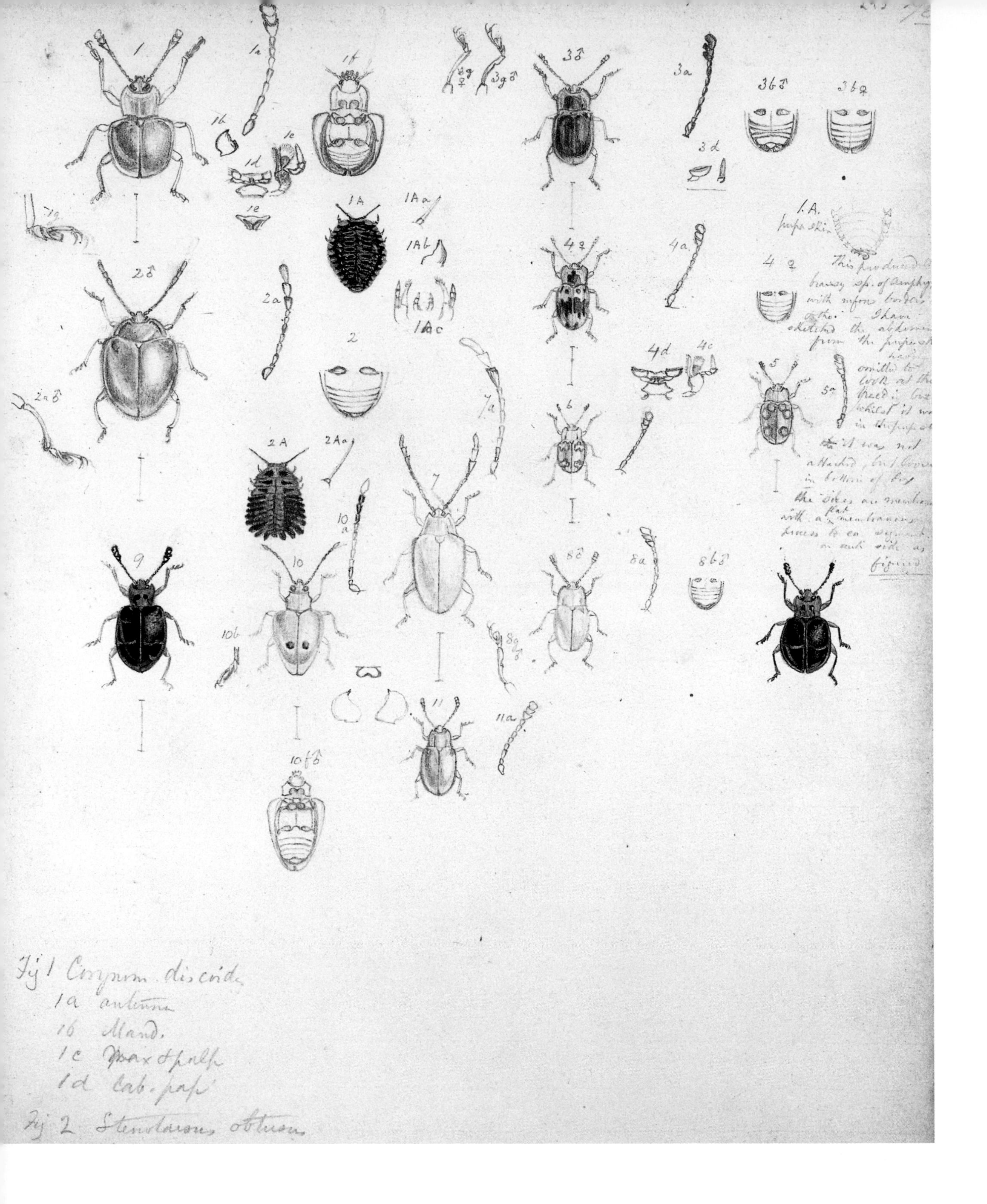
1a antenna
1b Mand.
1c Max & palp

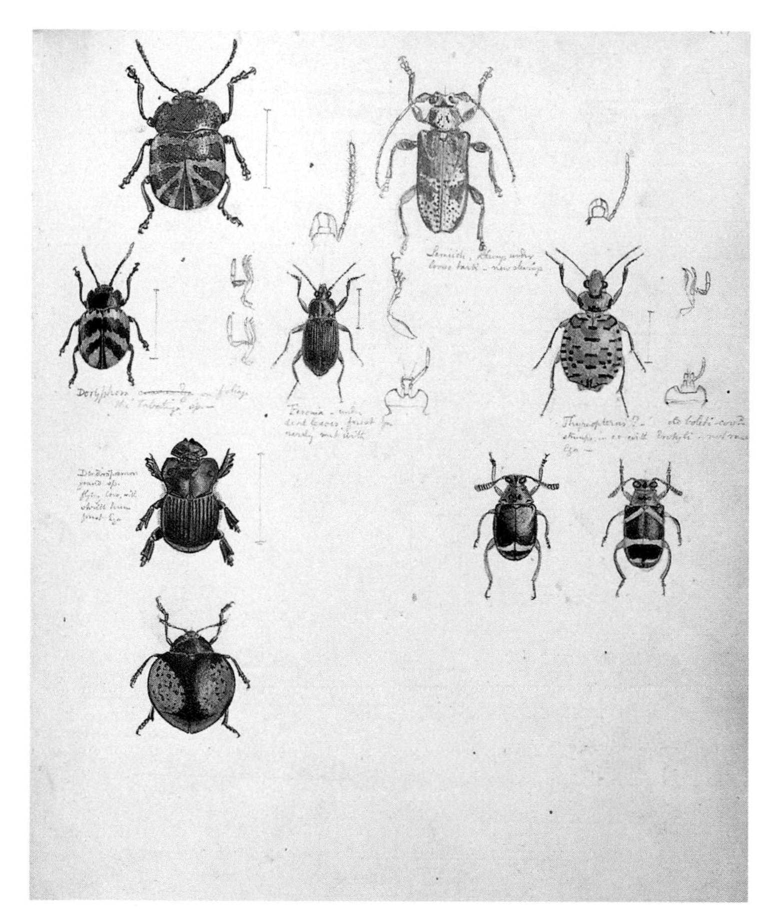

***Pages of Bates' beautiful beetle illustrations** (opposite, above & overleaf),* still as fresh and vibrant as the day he painted them over 150 years ago. These intricately annotated drawings represent a whole assortment of beetle families including the metallic carabids or ground beetles, scarabaeids and chrysomelid leaf beetles, and one longhorn beetle.

Prot[illegible]o.

mid tib

failing to mentum

Protho. cotyloid cav. opened behind

e
d
b
a
e

fore leg ♂

Protho.

Anisocerus
Allied to ~~[illegible]~~. colour, shape
body deceptively resemble the elongated
[A]nthribi .. on branches of fallen trees
[a]ntennae held close together on the
[b]ark. 4 Oct. 55

mid-tib

d

Megaderus Stigma. The lower lip
is composed of a very broad & short piece
a: of the same horny consistence as
the general integument: its upper edge
is cut out & joined to the membranous
piece b: forming the intermediate
piece between mentum & palpi –
~~[illegible]~~ c: is the paraglossae or ligula
or ligula & paraglossae united – it is
white cartilaginous & flexible & spring
(as I made quite sure) from the oesophagus
as far down as the base of mentum
it is one piece at its basal half &
soft, membranous or tumid – its upper
half is cleft. ~~&~~ within there is the
usual horny rib on each side running
up each of the lobes. Now it
appears to me that the ligula
here is reduced & invisible externally
the paraglossae being in recompense
highly developed.

The roots of lab. palpi are visible
– & soft. e. There is no trace of
the horny solidification of parts
as in Ctenoscelis No 36 – except
a small dark horny looking plate
at the bottom of the cleft of paraglossa
(d). This latter may be the remains
of the reduced ligula. 5 Oct. 55

♂ Anisocerus Onca, White B[illegible]

Allied to Lamia & especially to Omp[illegible]
cerus. The ♂ has fore tarsi 2 ba[illegible]
jts. fringed on sides with long hairs,
the apical jt. of ant. shorter than
the preced.g jts.

The labium is on same type
Megaderus & other Longicorn[s]
but the mentum, altho' horny
softer than the integument, & a[re]
coriaceous. The other parts a[re]
narrower & more elongate than
Megaderus. I see no trac[e]
of rib or keel on the inside of
the ligula-paraglos. 5 Oct. 55

The sp. is frequent at Ega in fo[rest]
on branches of fallen trees, often
found in cop.

The mandibles have not a to[oth]
w.i. apex as in Megaderus, but a[re]
faintly crenulated in the middle
inner side.

[H]ippopsis ? ♀ Antennae more
th[a]n 2ce the length of body. In ♂
[n]ot much longer, but the terminal jt.
longer. Prostern. dilated after
[the] coxa ~~[illegible]~~ & rather advanced
[t]owards the mesost. being broad &
[tr]uncated, there is a narrow slit
~~[illegible] between~~ [th]at wh. renders
[th]e cotyloid cav. not closed.
[T]he mentum as in figure. I
[n]oted particularly the horny piece
. it is elongate & narrow in
[th]is sp. & its tip furnished with
[a] stiff bristle exactly as the

[illegible] Fasenis

prost.

mesostern.

Chrysoprasis — the large, broad sp. middle lobe of maxillæ, greatly elongated, spoon shaped, like the Callichroma & unlike Trachyderes labium — ample expanded rounded lobes. 1st jt. of maxpalpi elongate, — mandibles toothed in middle — &c &c.

Ega 25 March 1856

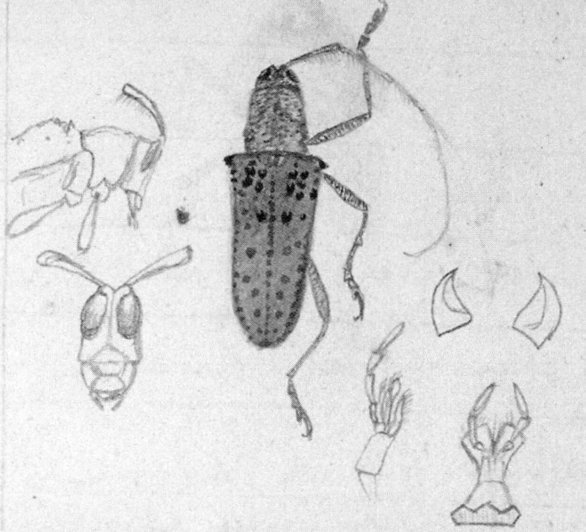

Lamiidæ Trachysomus — Mandibles broad blades, simply pointed not toothed — Front-plane elongate, rather narrow, eyes notched slightly at upper border antenniferous tubercles arising from the notch —

Ega 27 March 1856

Coremia hirtipes

Listroptera

Common small, wholly black — ½ elytra with blk stria, sternum surface rugose-corroded

Ega 21 June,

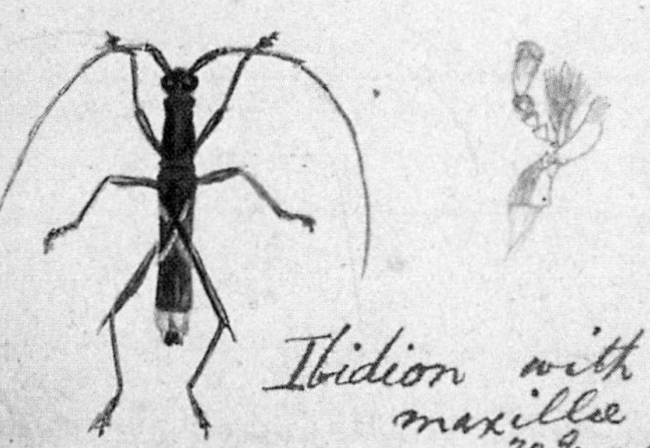

Ibidion with maxillæ 20 June 1856

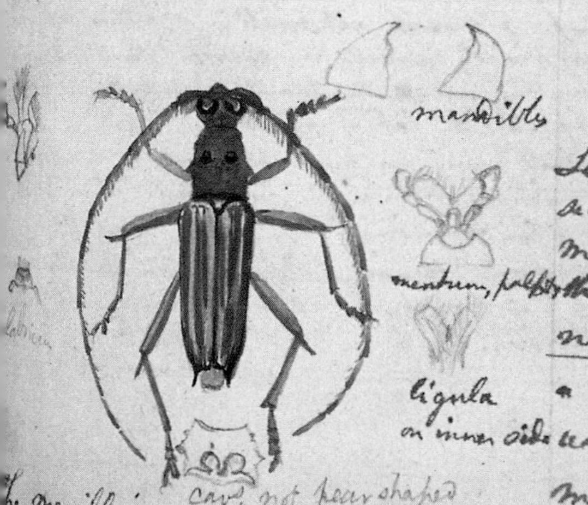

mandibles

mentum, palpi

ligula on inner side

cav. not pearshaped

Lamiidæ — nearest Leiopus — the second jt. of labial & the 2nd & 3rd of maxipalpi — much enlarged, tip jts thin & tapering — the Lingua is not cleft like other Longicornes, but a single piece, scarce even a sinuation in its upper edge. — The mandibles are suddenly narrowed

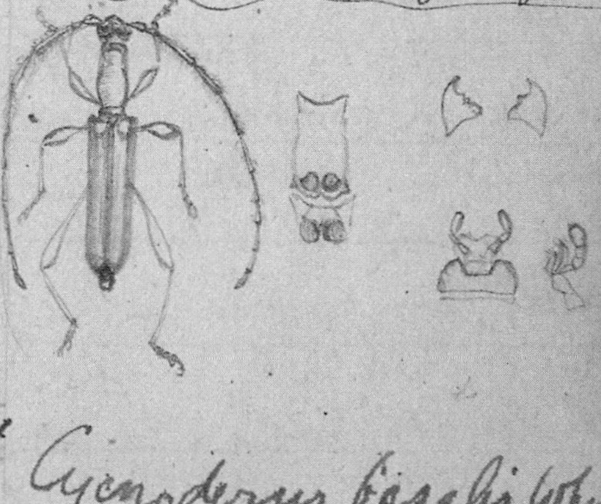

Cyenoderus basalis Whi

parts of the mouth, membranous or

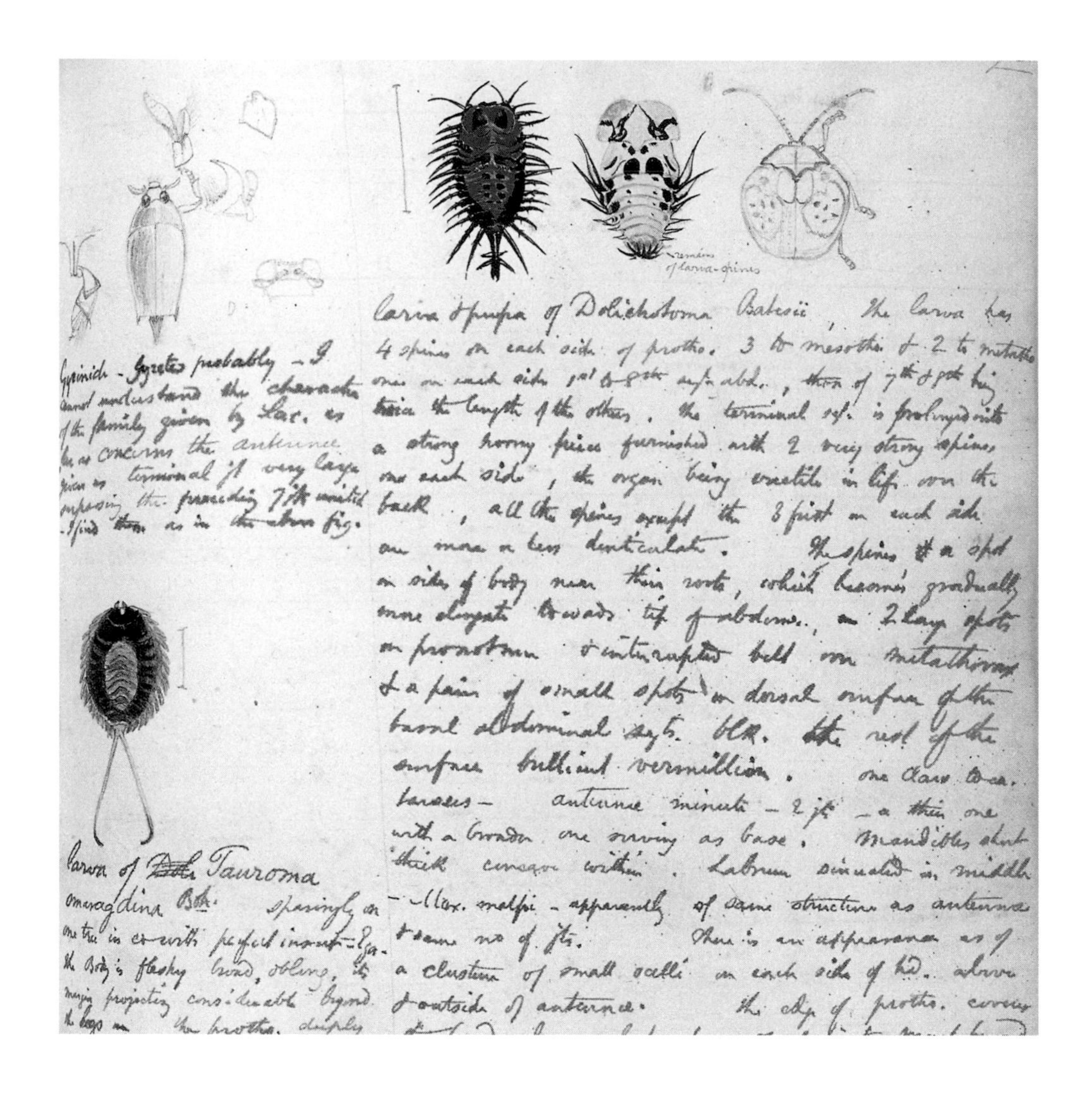

Scattered amongst Bates' meticulous notes *(above)* on the morphology, colour, ecology and behaviour of the insects are hundreds of his exquisite drawings. One page features illustrations of a curious family of grasshoppers *(opposite)*, unique to South America, which look remarkably like stick insects. In the accompanying notes, Bates points out that although the back legs are modified for jumping, they are much less well-developed than in more conventional grasshoppers so that these South American 'hoppers' are less mobile.

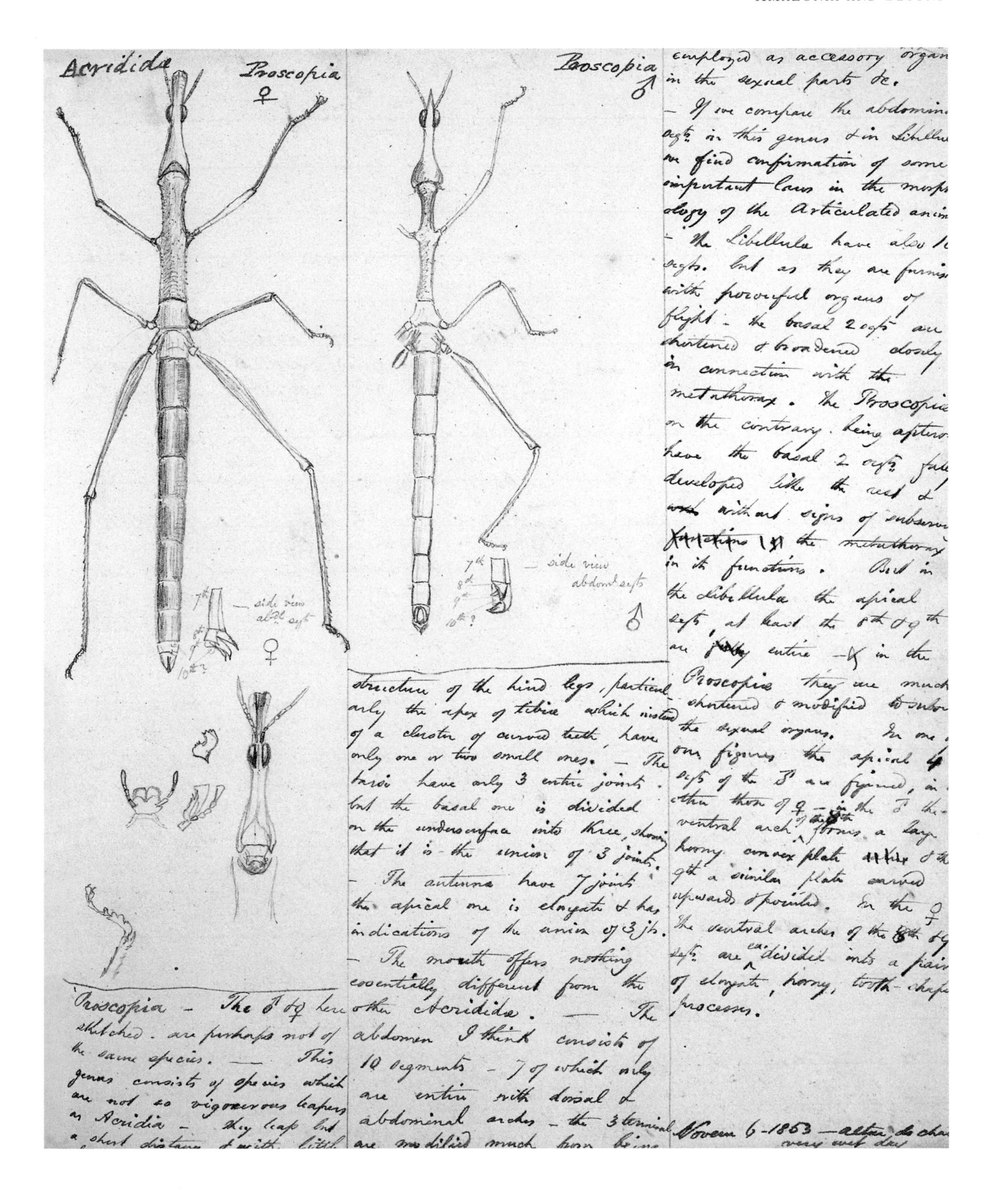

Acridida Proscopia ♀

Proscopia ♂

7th — side view abd^l segts

8th 9th 10th ?

7th 8th 9th 10th ? — side view abdom^l segts

Proscopia — The ♂ & ♀ here sketched — are perhaps not of the same species. — This genus consists of species which are not so vigorous leapers as Acridia — they leap but a short distance & with little

structure of the hind legs, particularly the apex of tibiae which consists of a cluster of curved teeth, have only one or two small ones. — The tarsi have only 3 entire joints. but the basal one is divided on the undersurface into three, showing that it is the union of 3 joints.

— The antennae have 7 joints the apical one is elongate & has indications of the union of 3 jts.

— The mouth offers nothing essentially different from the other Acridida. — The abdomen I think consists of 10 segments — 7 of which only are entire with dorsal & abdominal arches — the 3 terminal are modified much from being

employed as accessory organs in the sexual parts &c.

— If we compare the abdominal segts in this genus & in Libellula we find confirmation of some important laws in the morphology of the Articulated animals — the Libellulae have also 10 segts. but as they are furnished with powerful organs of flight — the basal 2 segts are shortened & broadened closely in connection with the metathorax. The Proscopia on the contrary being apterous have the basal 2 segts fully developed like the rest & without signs of subservience to the metathorax in its functions. But in the Libellula the apical segts, at least the 8th & 9th are entire — & in the Proscopia they are much shortened & modified to subserve the sexual organs. In one of our figures the apical segts of the ♂ are figured, in the other those of ♀ — in the ♂ the ventral arch of the 8th forms a large horny convex plate & the 9th a similar plate curved upwards & pointed. In the ♀ the ventral arches of the 8th & 9th segts are ea. divided into a pair of elongate, horny, tooth-shaped processes.

Novem 6 — 1853 — alter. do. — very wet day

A few of the many pages of Amazonian butterflies *(above & right)* from Bates' collecting notebook. The detail and colour in Bates' butterfly illustrations is impressive given the very basic facilities he had while out on his jungle expeditions. Some of the drawings were made directly into his notebook. Others were done on scraps of paper wherever he happened to be at the time and later pinned into the notebook.

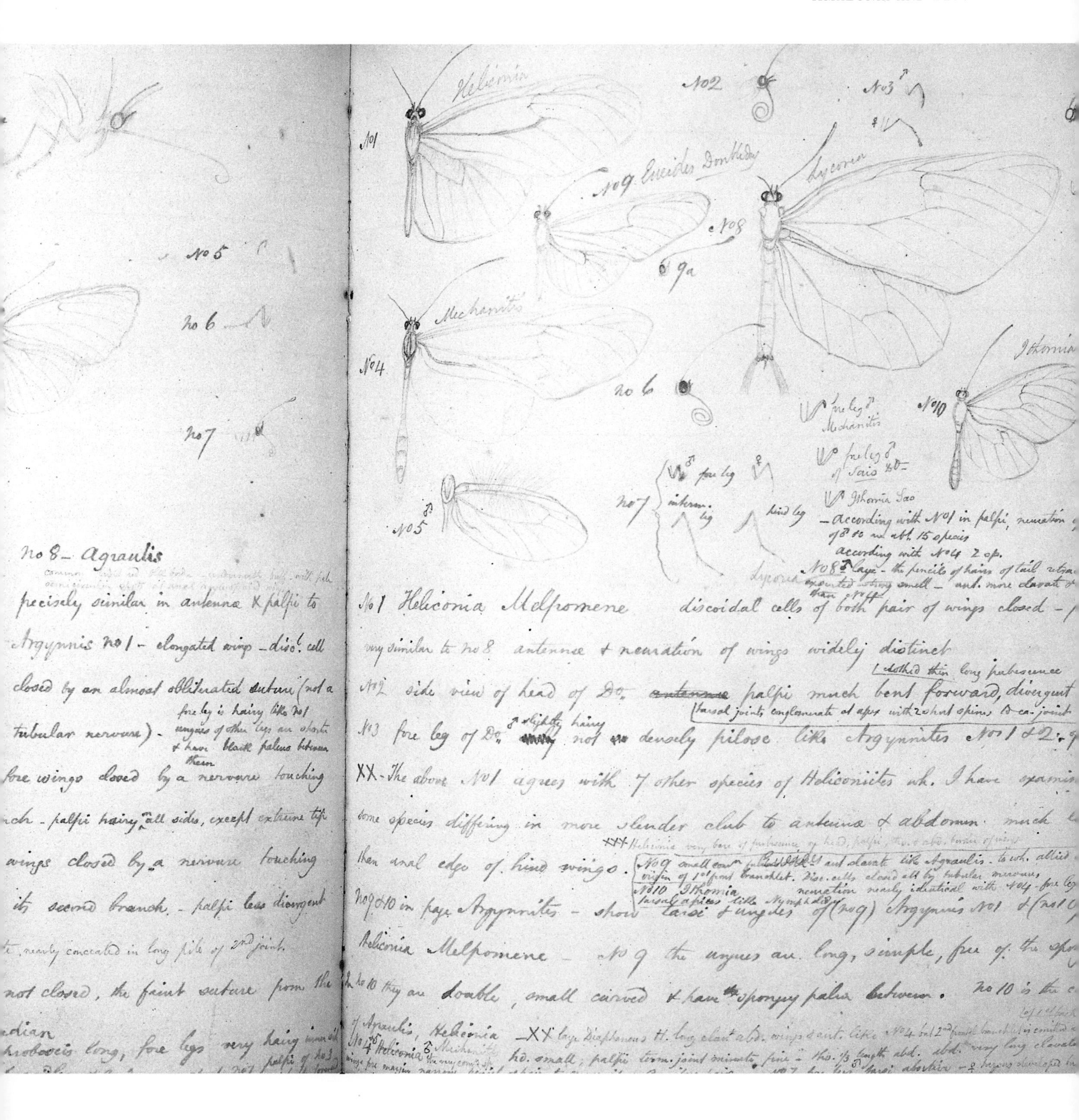
No 5
no 6
no 7

No 8 – Agraulis
pecisely similar in antennæ & palpi to
Argynnis No 1 – elongated wings – disc. cell
closed by an almost obliterated suture (not a
tubular nervure) –
fore wings closed by a nervure touching
nch – palpi hairy all sides, except extreme tip
wings closed by a nervure touching
its second branch. – palpi less divergent
not closed, the faint suture from the
dian
hrobosis long, fore legs very hairy

No 1
No 2
No 3
No 4
No 5
No 6
No 7
No 8
No 9
9a
No 10
Mechanitis
Lycorea
Ithomia

No 1 Heliconia Melpomene discoidal cells of both pair of wings closed –
very similar to No 8 antennæ & neuration of wings widely distinct
No 2 side view of head of Do. ~~antennæ~~ palpi much bent forward, divergent
No 3 fore leg of Do. not densely pilose like Argynnites No 1 & 2
XX – The above No 1 agrees with 7 other species of Heliconites wh. I have examin
some species differing in more slender club to antenna & abdomen much
than anal edge of hind wings.
No 9 & 10 in page Argynnites – show tarsi & ungues of (No 9) Argynnis No 1 & (No 10
Heliconia Melpomene – No 9 the ungues are long, simple, free of the spo
In No 10 they are double, small curved & have spongy palus between. No 10 is the c

, in ♂ they are

tion something

) in the forest

, wings

More butterflies from the Amazon *(opposite & above)*, which Bates not only sketched in his notebook but collected specimens of as well. In total, he collected 14,000 different species of insects, each of which had to be labelled and sent back to England. In his book he describes a typical working day in the jungle: 'I rose with the dawn and took a cup of coffee, and then sailed forth after birds. At ten I breakfasted, and devoted the hours from ten until three to entomology. The evening was occupied in preserving and storing my captures.'

Bates is revered not only for being methodical and scientifically astute in his approach to the butterflies and other insects he studied, but also for his artistic eye, as these arrangements of watercolour butterflies *(above & opposite)* make clear. Despite their aesthetic appeal, the specimens were of interest to collectors in London primarily for their scientific value, and as a demonstration of the great diversity present in the Amazonian region.

Fathoming the Deep 1872–1876

The *Challenger* Expedition

The larval stage of a sergestid shrimp *(opposite)* from Charles Spence Bate's *Challenger* report on crustaceans. Although technically an amateur zoologist, Spence Bate was highly professional in his scientific approach and drawing.

BY THE MIDDLE OF THE 19TH CENTURY ALL OF THE EARTH'S MAIN LANDMASSES, and most of the minor ones, had been 'discovered' and their coastlines, at least, pretty well surveyed. But despite more than two centuries of fairly intensive navigation all over the world's ocean by vessels from many nations, almost nothing was known of the nature of the deep seas beyond a depth of a few tens of metres. Even the basic depths of the great ocean basins were still a mystery, while it was commonly believed that the utter darkness, high pressures and intense cold assumed to characterise the deeper layers meant that they would be totally lifeless. In the 1850s and 1860s, however, a number of factors came together to change all this.

First, developments in submarine telegraphy and mounting pressure to lay cables between the continents generated a demand for accurate knowledge of the shape and nature of the deep-sea floor. Second, some tantalising, but inconclusive, new pieces of information suggested that the idea of a lifeless deep sea might be erroneous. Moreover, some of the animals found in early deep hauls were what we would now call 'living fossils' and adherents of the evolutionary process proposed in Charles Darwin's recently published *On the Origin of Species* thought that the deep ocean might hold many more. Clearly there were strong commercial and philosophical arguments for studying the deep oceans. And so it was that on 21 December 1872, a Royal Naval corvette, HMS *Challenger*, sailed from Portsmouth at the beginning of a three-and-a-half year voyage of scientific exploration which was to become the most famous expedition of its kind ever undertaken. It is often considered to represent the birth of oceanography, the science of the sea, for it was an absolute first. Never before had any nation despatched a major expedition with the express purpose of studying the physics, chemistry, geology and, particularly, biology of the deep ocean. At a total cost to the UK Treasury of almost £200,000 (well over £10 million today) it was also the world's first example of 'big science', for nothing like this amount had ever been spent on a single scientific undertaking before.

The voyage was the brainchild of two civilian biologists, Professor William Benjamin Carpenter (1813–85) from the University of London and Charles Wyville Thomson (1830–82), Professor of Natural History in the University of

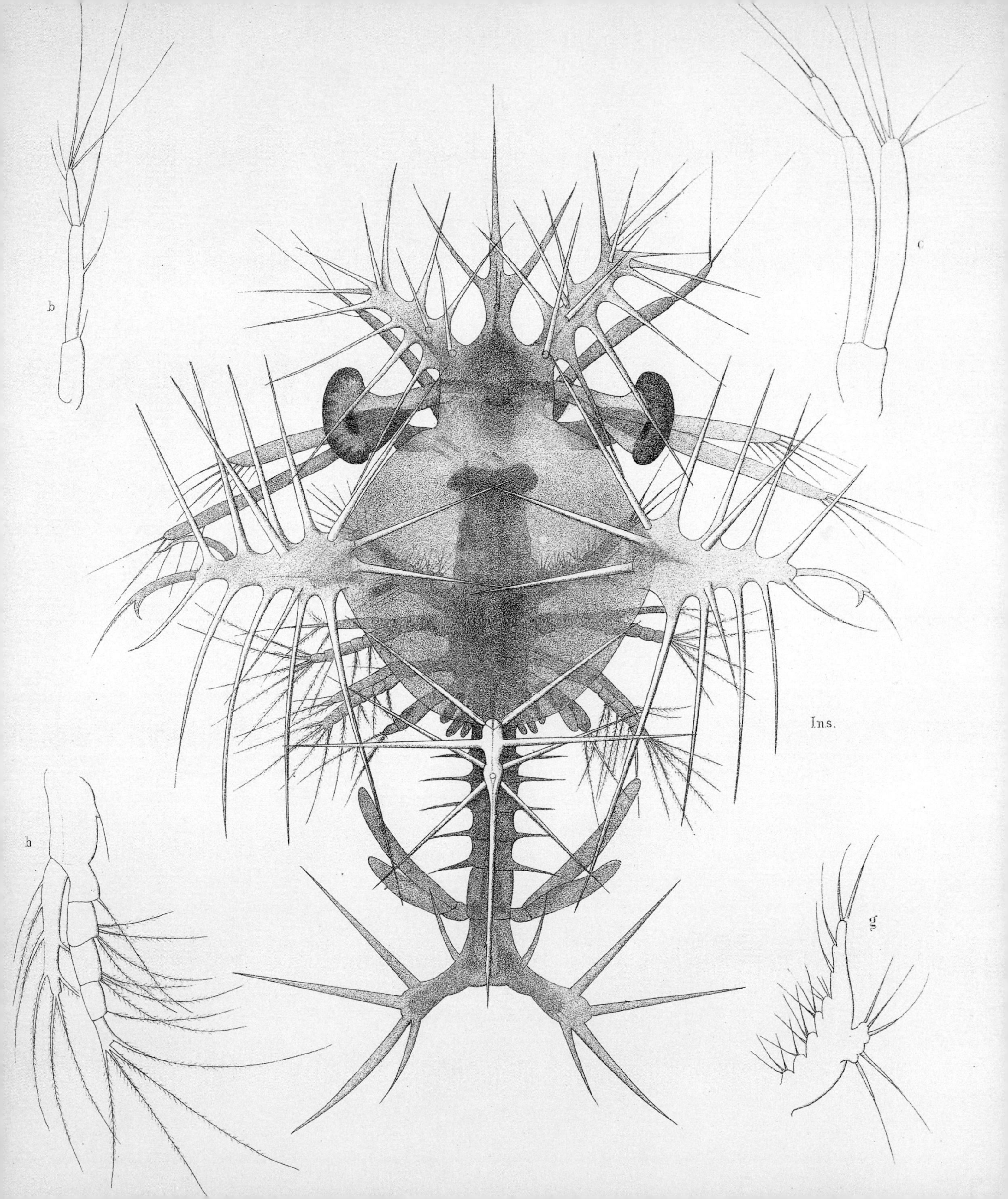
b
c
Ins.
h
g

HMS* Challenger *in the ice (opposite) in a painting by W. F. Mitchell. The ship spent two weeks in the Antarctic in February 1874. Despite the climatic extremes and dangerous waters traversed, only 10 of the original complement died, a remarkably good record for a voyage of that length and at that time.

Edinburgh. Like most other scientists of the day, until a few years before the *Challenger* Expedition Carpenter and Thomson had accepted the idea of a lifeless deep sea. But a series of short cruises on rather unsatisfactory naval ships around the British coasts between 1868 and 1870 had convinced them, correctly, that animal life would be found at all depths in the sea. These cruises also dispelled another myth, that the deep seas were universally filled with water at a temperature of about 4°C (40°F) – based on the erroneous assumption that, like fresh water, sea water has a maximum density at this temperature. As the preliminary cruises revealed, temperatures at the near-bottom of the ocean could be considerably lower or higher. But these results were based on a few observations in one corner of a single ocean. What was needed, Carpenter and Thomson reasoned, was a properly equipped global expedition to investigate the last great geographical unknown on earth. The audacious proposal was put to the Admiralty via the Royal Society and, with surprisingly little opposition, was accepted.

The submission had come at an auspicious time. Britain was at the peak of her imperial power; Britannia unquestionably ruled the waves and British jingoism was alive and well. Consequently, many within the Admiralty, and outside, thought that Britain should lead the way in any innovative maritime undertaking. Moreover, while the Admiralty Hydrographic Office had been the world's finest surveying and chart-making institution for several decades, until recently its interest in the deep ocean had been limited. But the submarine telegraph companies were asking unanswerable questions about the deep-sea floor so that quite apart from the scientific arguments, the proposal had the backing of the Hydrographer, Admiral G. H. Richards.

Within 18 months a ship had been selected, extensively refitted with purpose-built laboratories, civilian accommodation and winches, and supplied with tonnes of the latest equipment – including more than 400 kilometres (249 miles) of rope – for the extensive research programme. The ship's Captain was to be an experienced surveying officer, George Strong Nares (1831–1915), supported by a naval complement of about 225. The civilian scientific staff of six was headed by Thomson as chief scientist (at the age of 58, Carpenter had decided he was too old to participate) and included a

Scottish chemist, John Young Buchanan, three zoologists: one English, Henry Nottidge Moseley; one Scottish-Canadian, John Murray; one German, Rudolf von Willimöes-Suhm; and a Swiss artist, Jean Jacques Wild. This was a curiously appropriate mix for the beginnings of a discipline which has been characterised ever since by international collaboration.

After a short 'shake down' period to check the sampling techniques, and brief calls at Lisbon and Tenerife, the work began in earnest on 15 February 1873, some 64 kilometres (40 miles) south of the Canaries and at a depth of 3,500 metres (11,483 feet). Here, the *Challenger* personnel worked what would be the first of many official 'stations' or study sites dotting the world's oceans. By the time the ship returned to Spithead on Queen Victoria's 51st birthday on 21 May 1876, the *Challenger* had criss-crossed all of the major oceans other than the Indian, covering a total of 68,890 nautical miles, mostly under sail despite her 400 nominal horsepower coal-fired steam engine, coupled to a twin-blade propeller. She had spent more than half of the intervening days in harbour, providing her sailors and scientists with the opportunity for exotic port calls in north and south America, South Africa, Australia, New Zealand, Hong Kong, Japan and a series of Atlantic and Pacific islands. During these runs ashore the ship's officers and scientists made extensive zoological, botanical and ethnographic collections, and met all kinds of people, ranging from the King of Portugal and the Emperor of Japan to Fijian natives who had only recently given up cannibalism. Many of these events and views were documented by the official artist, Jean Jacques Wild. But many were also recorded photographically, because the *Challenger* Expedition seems to have been the first to routinely use this relatively new technology. Photography was, of course, still in its infancy and the long exposure times required made it quite unsuitable for action shots, for example of the scientists or sailors working on deck. But it was excellent for recording individuals or groups of people as well as distant views; indeed, one of the requirements in the expedition's official instructions was that: 'Every opportunity should be taken of obtaining photographs of native races to one scale.' Consequently, the resulting superb collection of photographs provides an historically important record of places and people in the late-19th century – including, probably the very first photographs of Antarctic icebergs ever taken.

But the *Challenger*'s main work, of course, had been accomplished during her 713 days at sea when, every two or three days on average, she would stop and work one of her 362 official stations. At each one, the depth of water was measured and a small sample of the bottom sediment was collected. The water temperature at the surface, near the bottom and often at several intermediate depths was measured and water samples collected for later chemical analysis. Finally, biological samples were obtained, always using a dredge or trawl dragged across the seabed and often also plankton nets to collect midwater animals down to about 1,500 metres (4,921 feet). Each time, the catch had to be carefully sorted, preserved, bottled, labelled, stored and meticulously documented. The speed and direction of the surface currents were also recorded fairly routinely while attempts to measure the shallow subsurface currents were made rather less regularly.

During the course of the voyage, material was sent back from Bermuda, Halifax, the Cape, Sydney, Hong Kong and Japan. In his introduction to the scientific reports Thomson wrote that '... after the contents of the ship had been finally cleared out at Sheerness, we found, on mustering our stores, that they consisted of 563 cases, containing 2,270 large glass jars with specimens in spirit of wine, 1,749 smaller stoppered bottles, 1,860 glass tubes, and 176 tin cases, all with specimens in spirit; 180 tin cases with dried specimens; and 22 casks with specimens in brine.' He adds that '... of upwards of 5,000 bottles and jars of different sizes sent from all parts of the world to be stored in Edinburgh, only about four were broken, and no specimens were lost from the spirit giving way.' Despite the importance of these specimens, the mechanics of collecting, storing and recording them was tedious, as was the cataloguing of the endless measurements made at each of the stations. Even for most of the scientists the novelty wore off after the first few tens of stations, while several of the officers documented their boredom in their personal journals. It must have been much worse for the ordinary sailors who had to do all the hard physical work without the reward of a vested interest in the results. No wonder 61 of them deserted at the various port calls.

But it was the sheer routine that was the strength of the *Challenger* Expedition. Almost nothing accomplished from the ship was absolutely new, for the techniques employed were mainly tried and tested ones which had been used

The **Challenger** ***photographers*** used a wet plate camera, similar to this one *(opposite)*, to take numerous photographs of landscapes, peoples and objects of interest over the duration of the expedition. The technology was not yet advanced enough, however, to be of any great use in scientific recording.

sporadically by ships of many nations. The significance of the *Challenger* was the intensity of the observations made, the global coverage and particularly the emphasis on very deep water. Her deepest sounding, at almost 8,200 metres (26,904 feet) in what later became known as the Challenger Deep in the south-western Pacific, was by far the deepest taken up to that time, at a site very close to where the current record depth – of a little over 11 kilometres (seven miles) – was measured. Similarly, her 25 successful dredgings at depths in excess of 4,500 metres (14,765 feet), and the deepest at 5,700 metres (18,701 feet), were totally unprecedented.

The resulting vast biological collections were shipped back periodically to Edinburgh to await the ship's return. For like all scientific undertakings, it was realised that the collections and the data would be of little value until they had been intensively studied and the results published. Accordingly, Thomson established a *Challenger* office in Edinburgh after the voyage to collate all the data, despatch the biological specimens to specialist scientists and supervise the publication of the resulting reports. This was to take much longer than the voyage itself, and after Thomson's death in 1882 the supervisory role was taken over by John Murray (1841–1914), one of the junior naturalists on the voyage but destined to become one of the most famous scientists of his day. It also turned out to be a much bigger task than had been originally envisaged. Thomson had estimated that the results would run to about 15 volumes to be published by the Stationery Office within about five years. In the event, the official reports ran to 50 large volumes with a total of 29,552 pages. The final two volumes appeared in 1895, 19 years after the end of the expedition.

The preparation of the reports had also involved Thomson and subsequently Murray in a series of rows. Firstly, this was because the authorities of the British Museum thought they should receive the collections and organise the work on them, rather than Thomson in Edinburgh. Many also thought that only British scientists should work on the material, rather than the best scientists, irrespective of nationality, as Thomson wished. Thomson won, and the collections were sent to an international galaxy of experts from France, Germany, Italy, Belgium, Scandinavia and the United States, as well as from the UK. The specimens eventually went to The Natural History Museum in London, where they remain to this day.

Scientific dignitaries from the Royal Society together with some of the expedition's scientists *(opposite)* on board the *Challenger* at Sheerness in early December 1872 before she sailed round the coast to Portsmouth for the final departure on 21 December. Some of these gentlemen, including the expedition's scientific leader Thomson, in the centre of the middle row, did not keep their heads still long enough for the slow emulsions of the photography of the day.

But the longest and most difficult battle was with the Treasury which was most unwilling to foot the increasingly large bill for the publication of the reports. This undoubtedly contributed to Thomson's failing health and his death at the early age of only 52. Murray was eventually able to secure the Treasury's financial backing, partly by shaming them when he threatened to finance the publication himself.

The final result is that the excellent reports, residing in the libraries of dozens of major laboratories around the world, are constantly referred to by modern oceanographers. And rarely a day goes by when a visiting scientist from one or other of those laboratories is not sitting in a room in The Natural History Museum, carefully studying specimens from the superb collections on which the reports are based. The reports contain thousands of illustrations, including many photographs and a few of Wild's watercolours. But Wild was only moderately talented as an artist, and although he illustrated some of the animals collected in their fresh state, the vast majority of the originals of the illustrations in the reports were produced either by the individual scientists to whom the various animal groups were sent, or by artists and engravers engaged by them. Consequently, instead of being dominated by one or a small number of artists, the pictorial material from the *Challenger* Expedition is the work of literally dozens of artists, engravers and lithographers most of whom never even saw the ship let alone sailed on her.

Despite the accolades which the *Challenger* Expedition has justifiably received from marine scientists in the subsequent century and a quarter, it did not encourage the British Government to continue support for oceanography in the immediate aftermath. So shocked was the Treasury at its unintended generosity, that it was to be several decades before it again became involved in a comparably expensive scientific undertaking. Fortunately, other nations did follow the *Challenger's* lead. In the last quarter of the 19th century, major oceanographic expeditions were despatched from the United States, Germany, Norway, Sweden, France, Italy and Monaco. The newly opened oceanographic trail blazed by the *Challenger* was well on its way to becoming the broad international co-operative highway that it is today.

GOV. COPYT.
A. J.H.

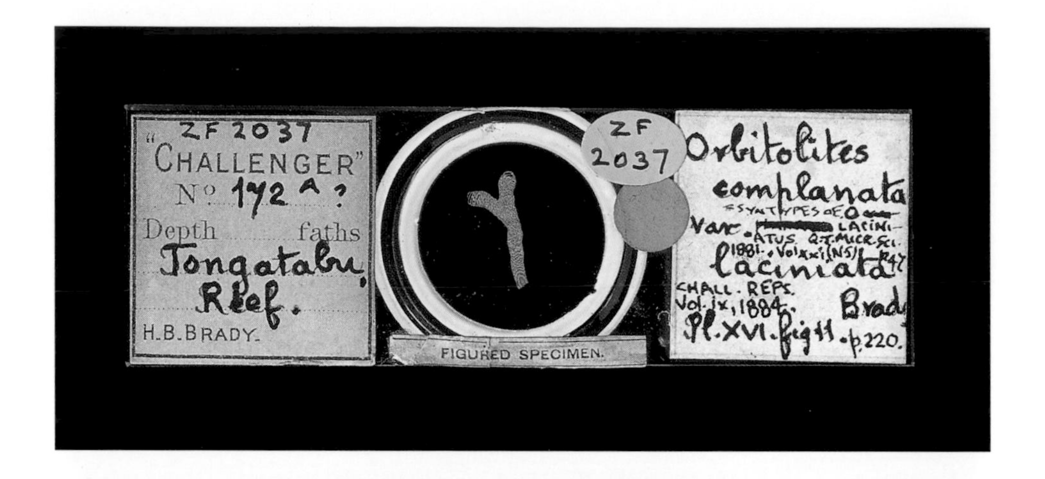
ZF 2037
"CHALLENGER"
No 172 A ?
Depth faths
Tongatabu Reef.
H.B.BRADY.
FIGURED SPECIMEN.
ZF 2037
Orbitolites complanata
laciniata
CHALL. REPS.
Vol. ix, 1884. Brady
Pl. XVI. fig 11. p. 220.

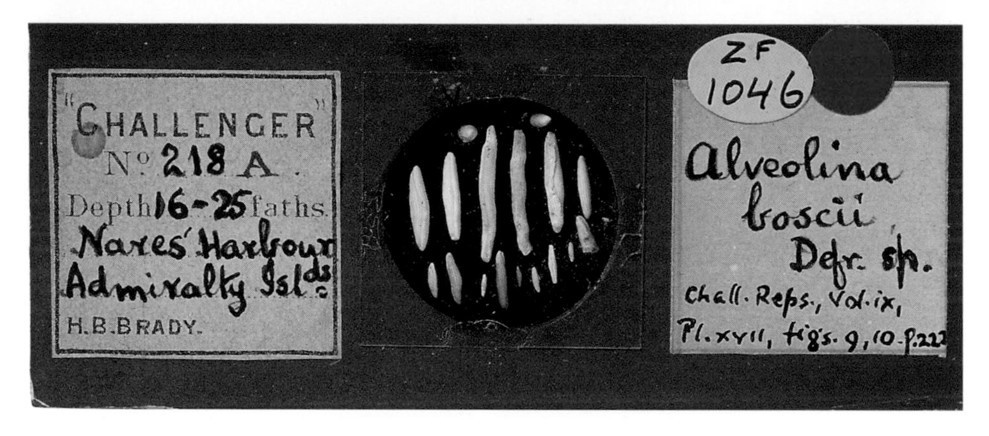
ZF 1046
"CHALLENGER"
No 218 A.
Depth 16-25 faths.
Nares' Harbour
Admiralty Islds
H.B.BRADY.
Alveolina boscii
Defr. sp.
Chall. Reps., Vol. ix,
Pl. XVII, figs. 9, 10 - p. 222

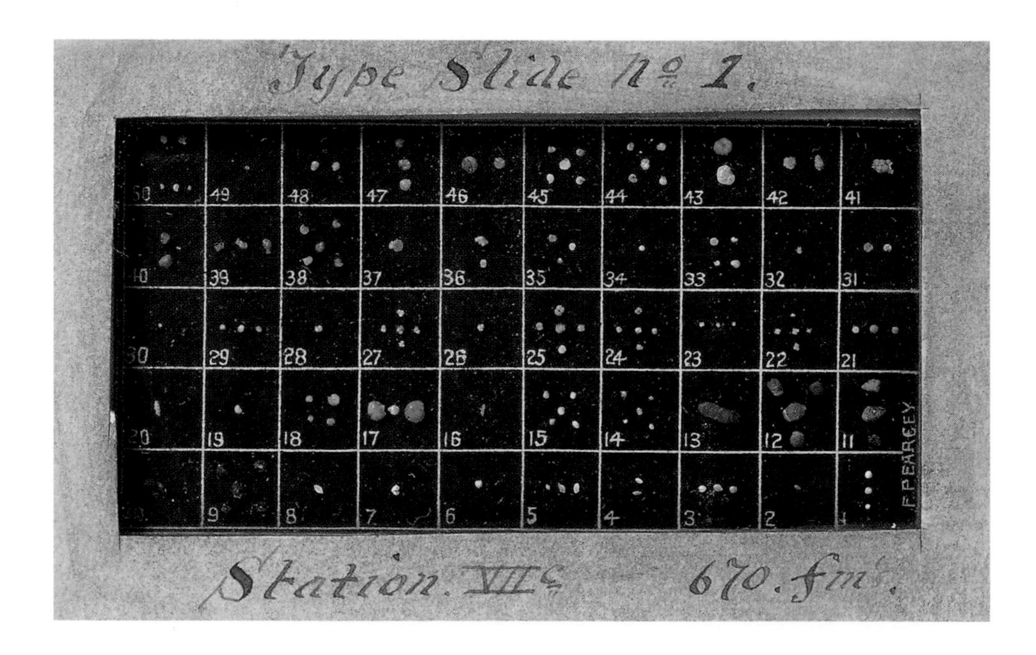
Type Slide No 1.
F.P.PEARCEY
Station. VII 670. fms.

An engraving of the* Challenger*'s laboratory *(above)* shows the work benches, microscopes, bottle racks and bird skins hung up to dry. The 'dirty' work associated with the net catches when they first came aboard was mostly done on deck or in a shed which was built at the aft end of the main deck. When the ship reached warm latitudes the drying bird skins were also hung in this 'shed' because the smell below decks was otherwise intolerable.

In addition to the multicellular animals brought up by the *Challenger's* trawls and dredges, vast numbers of unicellular organisms, particularly foraminifera, radiolaria and diatoms, were also amassed, many of them belonging to previously undescribed species. The foraminifera, whose dead shells make up the bulk of deep-sea sediments, were studied by Henry Bowman Brady, who meticulously sorted the dried mud samples and mounted tiny individual specimens on slides *(opposite),* now in the collections of the Palaeontology Department of The Natural History Museum.

The smoking crater of the active volcano Kilauea in Hawaii *(above)*. The *Challenger* reached the Hawaiian Islands in August 1875 and one of the high points of the visit was a trip to Kilauea, some 1,219 metres (4,000 feet) above sea level and involving a long and tedious journey from the harbour at Hilo. But the *Challenger* personnel were surprised to find a quite reasonable hotel already established on the rim of the crater to accommodate visiting tourists. At the other extreme, in February of the year before, the *Challenger* spent two weeks amongst the Antarctic ice, during which time icebergs were photographed from the ship *(opposite)*. This is one of the first photographs ever taken of Antarctic icebergs. (The wave-like pattern at the bottom is not water but was caused by the chemical masking agent used in development.)

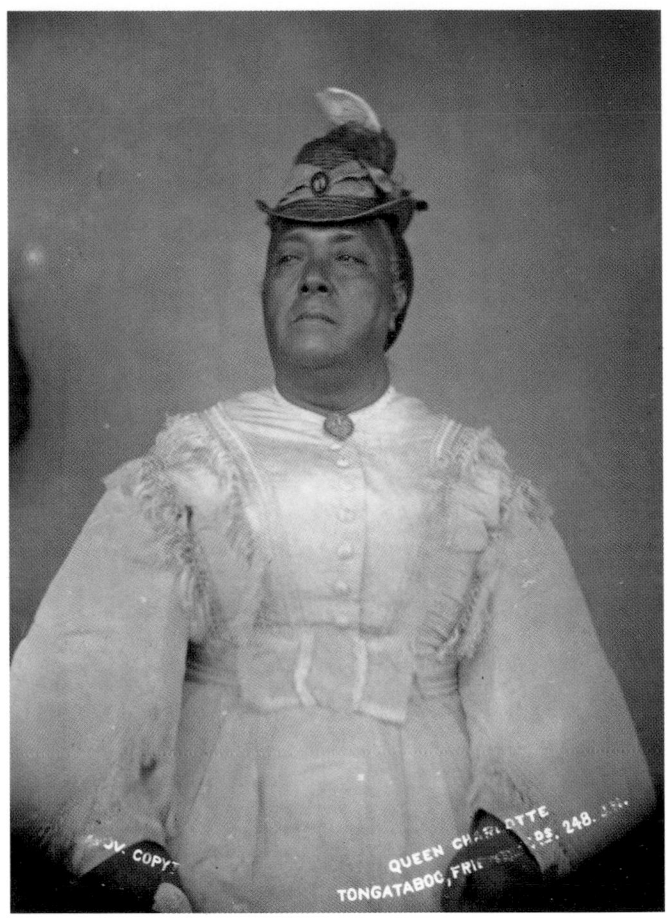

Photography in the 1870s was ideal for carefully posed portraits and the *Challenger* photographers produced many, like these of a Moro Indian from Zamboanga, Mindanao *(above left)*, a Japanese man *(opposite)* and Queen Charlotte of Tonga *(above right)*. The latter is one of an impressive collection of royal portraits made during the voyage, including King Luiz I of Portugal, Emperor Matsuhito of Japan and King Kalakaua of Hawaii. Queen Charlotte and her husband King George Tupou were particularly anxious to be photographed, the King in naval uniform and Charlotte in 'a light muslin costume of European make'. Her great, great grand-daughter, Queen Salote, bore a strong family resemblance and almost stole the limelight when she attended the coronation of Queen Elizabeth II in 1953.

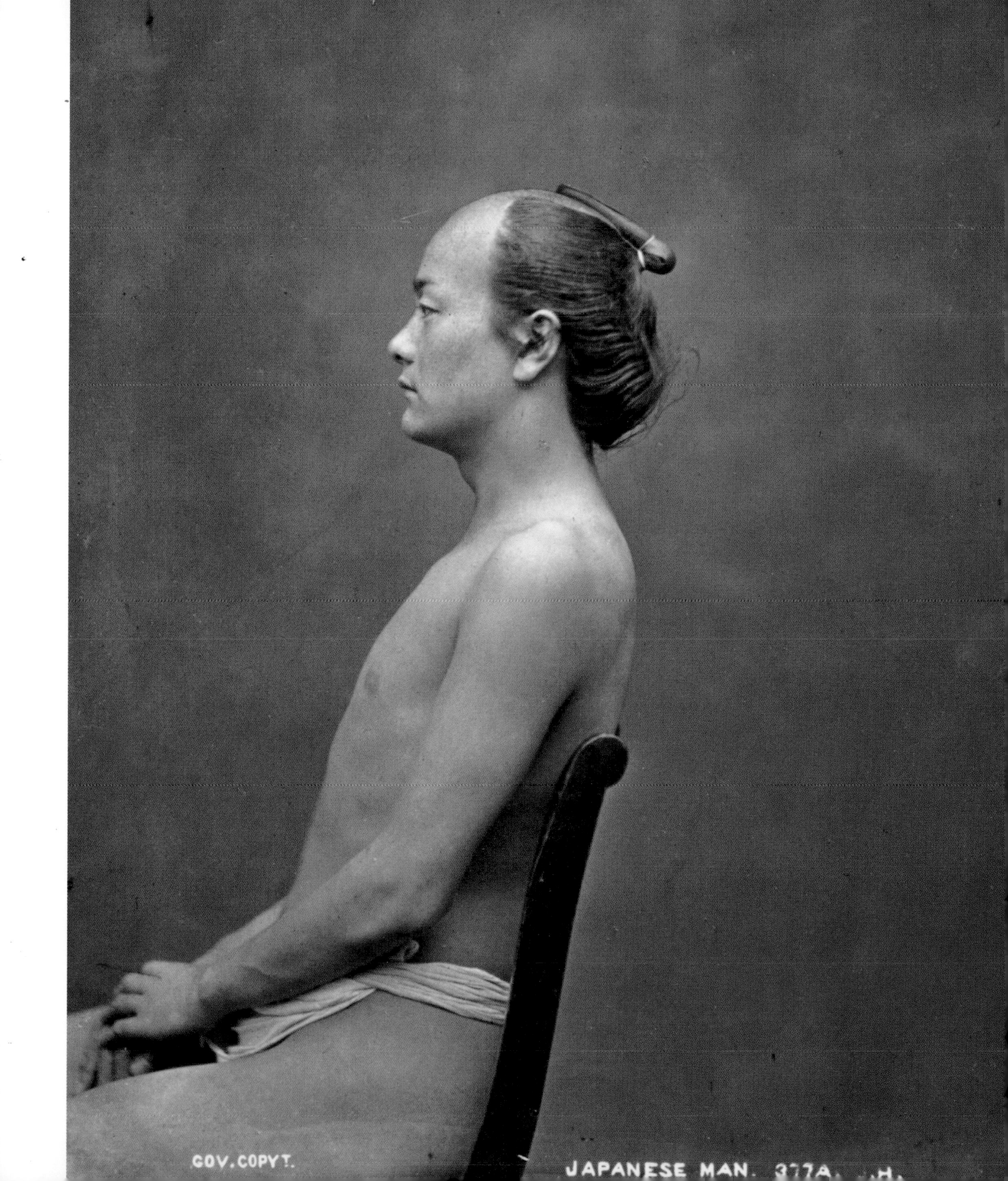
GOV.COPYT.
JAPANESE MAN. 377A. .H.

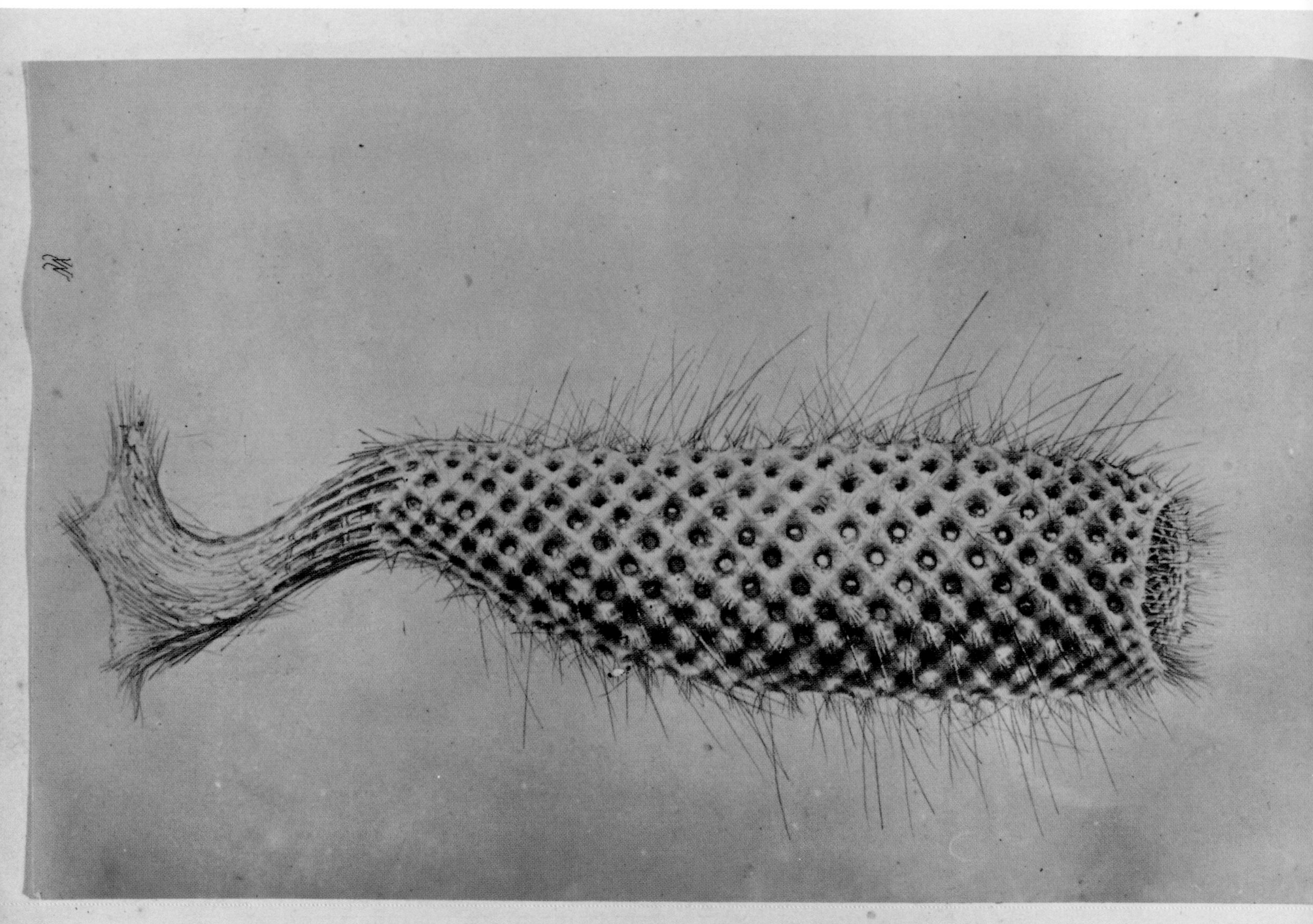

Sponge. (Euplectella.)

Drawing by Jean Jacques Wild of a glass sponge, probably *Euplectella suberea (opposite)*. This drawing would have been 'reconstructed' from several damaged specimens collected by the *Challenger* in deep water to the west of Gibraltar, and between Pernambuco and Bahia in South America. Wild also drew the deep-ocean sea-pen *Umbellula thomsoni (above)*, named in honour of the *Challenger* Expedition's scientific leader, Charles Wyville Thomson. These impressive plant-like animals are related to sea anemones, corals and jellyfish. They stand rooted in the seabed sediments and their stalks support a crown of filtering cells, which collect food particles from the passing water currents.

***The goosefish,* Lophius naresii**, now *Lophiodes naresi (opposite)* from Albert Gunther's report on the shallow-water fishes collected during the *Challenger* Expedition. This one, now known to be distributed in the south-west Pacific and south-east Indian Oceans, was undescribed at the time and was named in honour of the *Challenger*'s Captain, George S. Nares. It is related to the angler fish, monk or lotte of European waters. The type specimen of the goosefish *(above)* remains in good condition in the preserving jar to which it was consigned in the 1870s.

The Voyage of H.M.S. "Challenger"
Deep-sea Fishes. Pl. XI.
D.

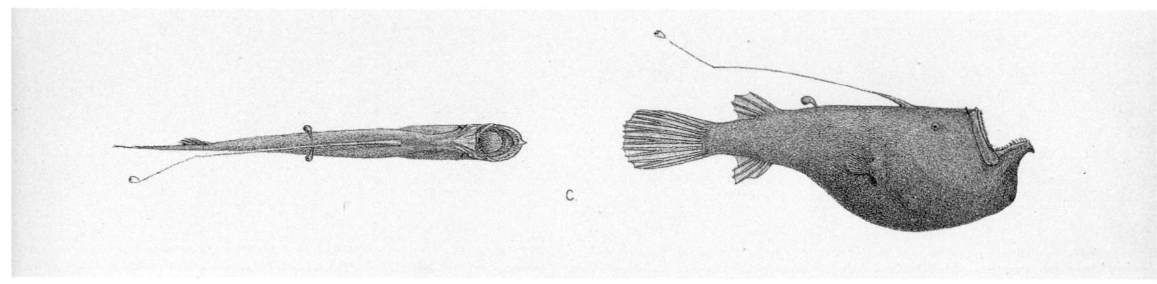
C.

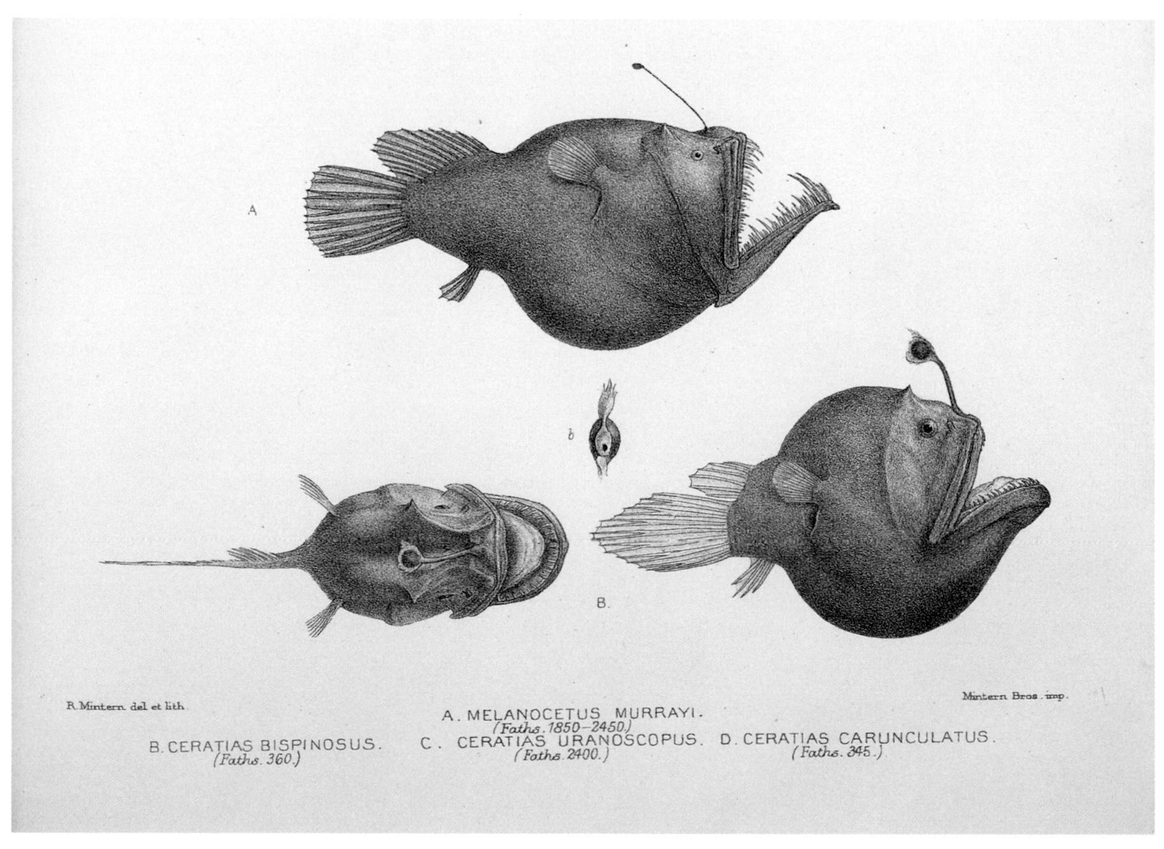
A
b
B.
R. Mintern del et lith.
Mintern Bros. imp.
A. MELANOCETUS MURRAYI.
(Faths. 1850–2450.)
B. CERATIAS BISPINOSUS.
(Faths. 360.)
C. CERATIAS URANOSCOPUS.
(Faths. 2400.)
D. CERATIAS CARUNCULATUS.
(Faths. 345.)

***Deep-sea fish** (opposite)* from the *Challenger* reports, including *Melanocetus murrayi* in the centre, named after John Murray, one of the expedition's junior zoologists and ultimately the most famous of the *Challenger* scientists. The reports also feature shore fish *(above)*. From the top is: the threadfin filefish, *Monacanthus* (now *Paramonacanthus*) *filcauda*, from the Arafura Sea; the tessellated filefish, *Monacanthus* (now *Thamnaconus*) *tessalatus*, from the Philippines; the lizardfish, *Saurus* (now *Synodus*) *kaianus*, also from the Arafura Sea; and the gaper or sabregill, *Champsodon vorax*, distributed in the eastern Indian and West Pacific area.

Several plates in the sea urchin volume of the *Challenger* reports are devoted to the details of the morphology of the animals' spines. This one *(opposite)* shows spine sections cut for Alexander Agassiz of the Museum for Comparative Zoology, Harvard, and drawn and engraved by James H. Blake, one of the many artists whose work graces the reports. Two southern Australian fishes *(below)* from Gunther's shallow water fish report: At the top is the Peacock Skate, *Raja* (now *Pavoraja*) *nitida,* and below is the Spiny Pipehorse, a sort of intermediate between a pipefish and a seahorse, *Solenognathus fasciatus* (now *spinosissimus*), known from shallow muddy waters around south-eastern Australia, Tasmania and New Zealand.

19.
3.
5.
13.
4.
6.
2.
8.
9.
14.
12.
22.
21
37.
24.
16.
33.
38.
17.
28.
36.
34.
30
26.
31.
32.
35.
25.
27.

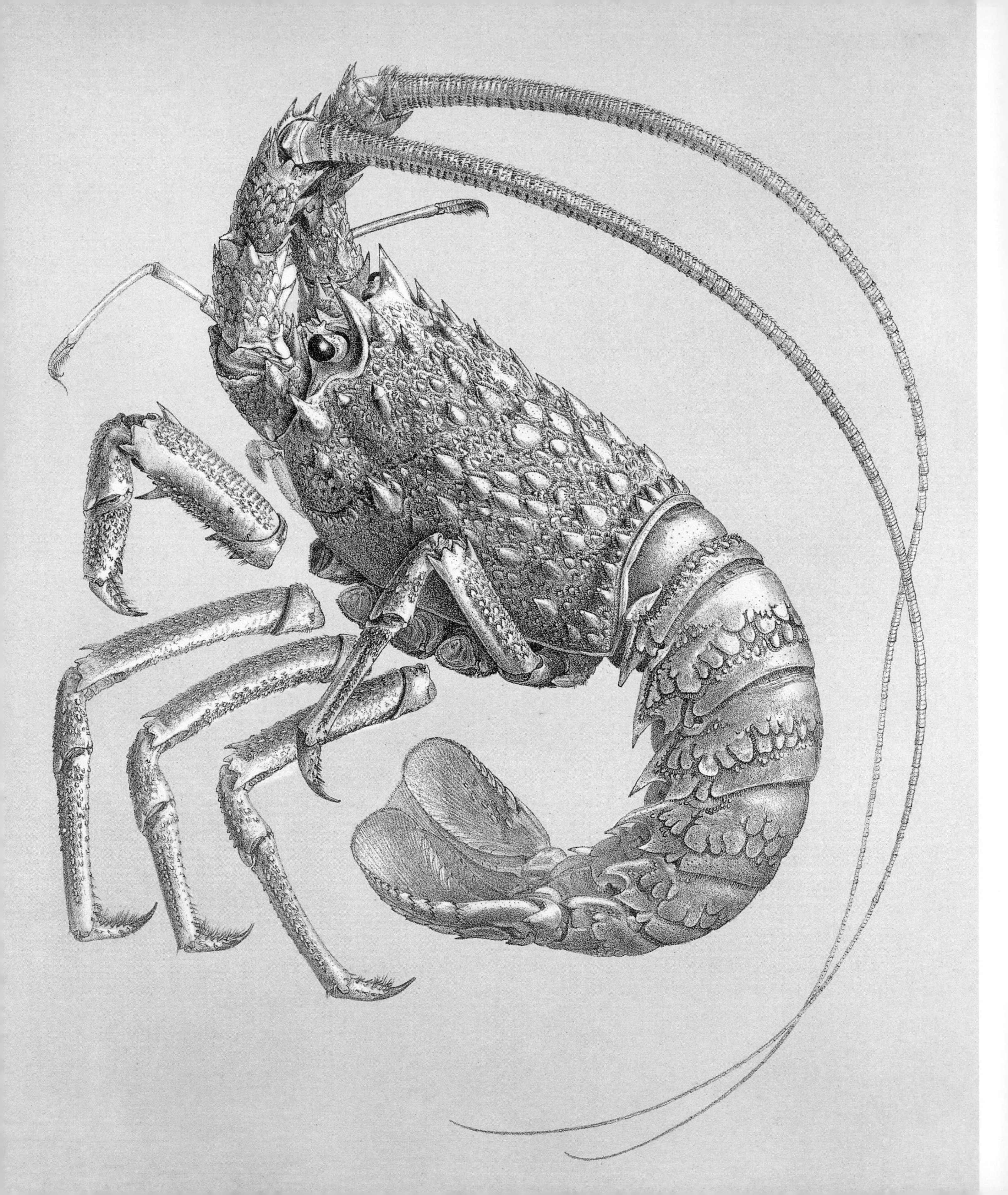

From Charles Spence Bate's report on long-tailed crustaceans are *Amphion provocatoris (below left), Amphion reynaudii (below right)* and the spiny lobster *(opposite)*, collected from Tristan da Cunha. Spence Bate decided that the lobster belonged to a species described by French zoologist, Henri Milne Edwards, as *Palinurus lalandii*, but thought it was sufficiently different from the other species to create a new genus, *Palinostus*. Spence Bate's new genus name was subsequently replaced by the older name *Jasus*, so that the full name for Milne Edwards' species is now *Jasus lalandii*. However, the *Challenger* specimen did not in fact belong to *lalandii*, but to a species not yet described at the time. In 1963 a Dutch crustacean expert recognised that spiny lobsters from Tristan da Cunha, including the *Challenger* specimen, belong to a distinct species which is now called *Jasus tristani*.

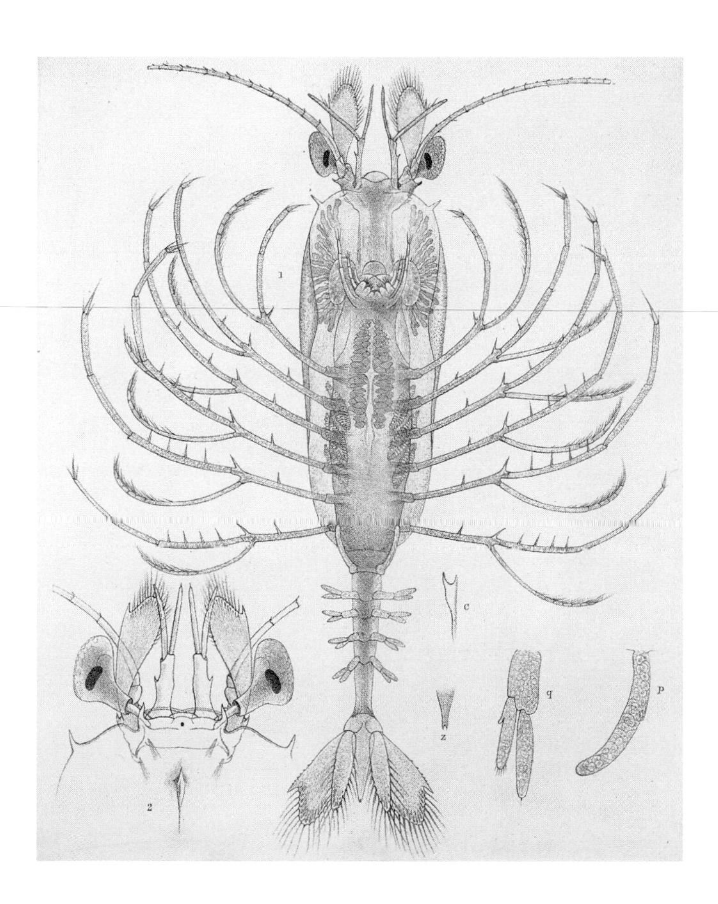

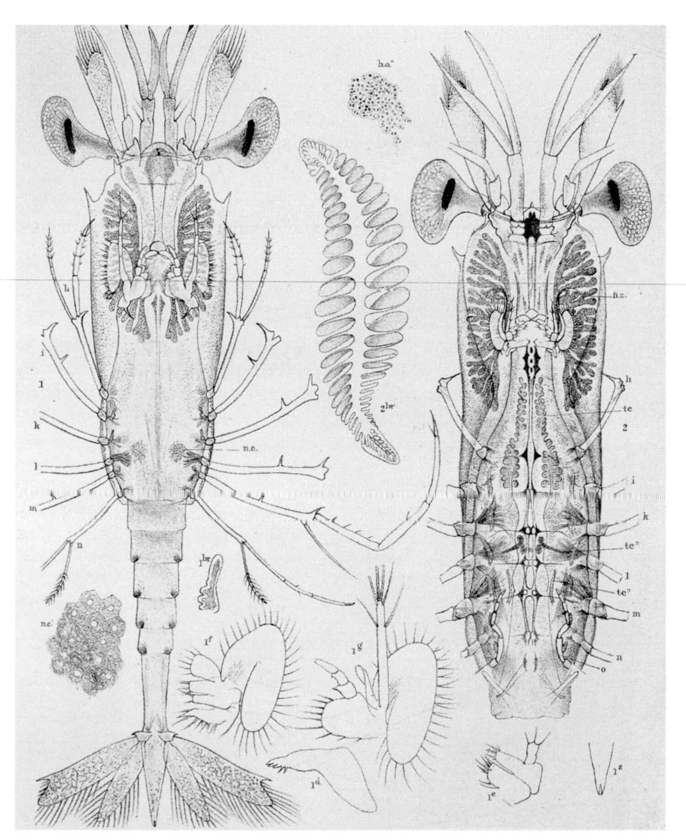

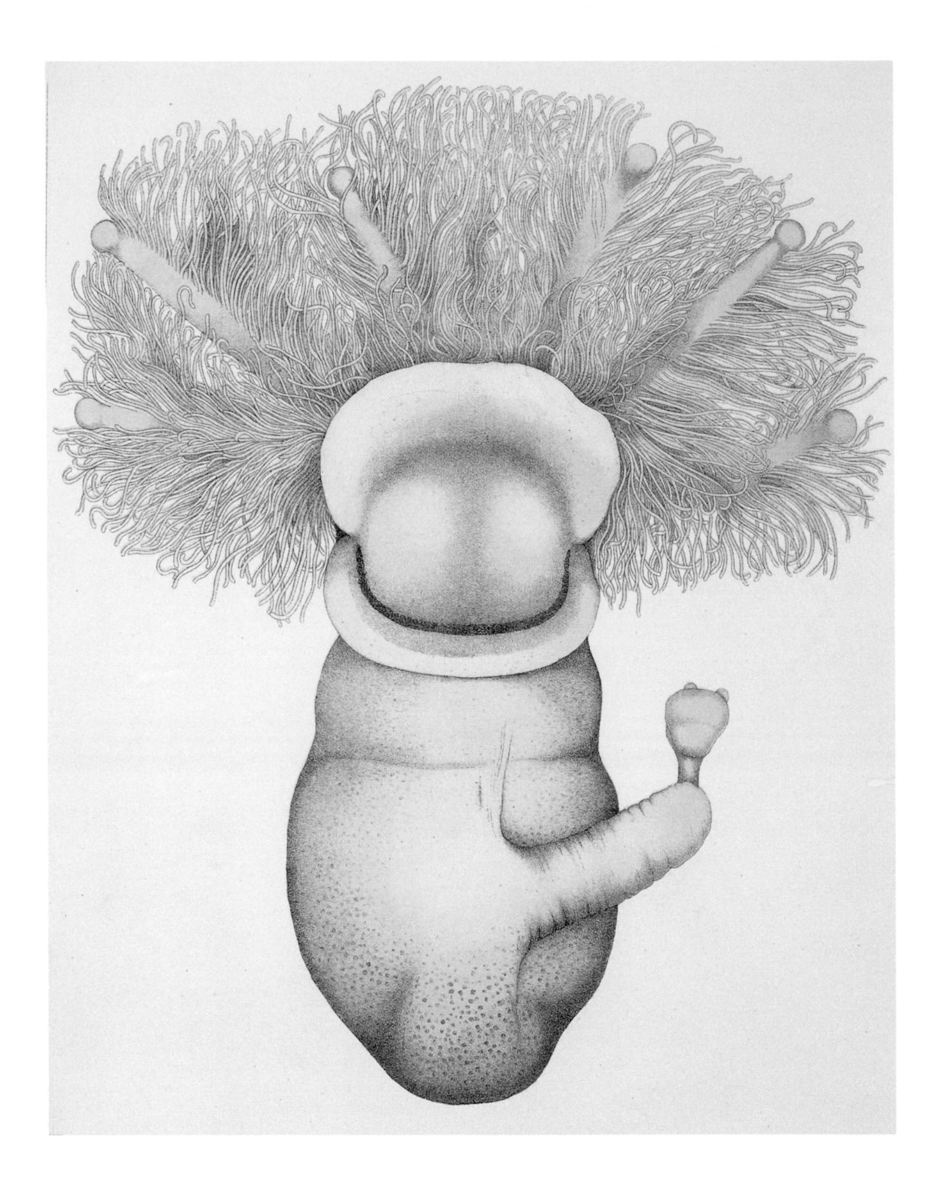

A tiny enteropneust, **Cephalodiscus dodecalophus** *(above),* which belongs to the protochordates – a group of animals which, like sea squirts, are intermediate in form between those with backbones (the vertebrates) and those without backbones (the invertebrates). It collects food using twelve tentacular plumes, hence its species name, though only six of them are visible here. In contrast to its soft contours are the spiky cidarid sea urchins *(opposite)* collected by the *Challenger* and rendered directly as a lithograph by Paulus Roetter. This is one of 45 plates illustrating the official report on the *Challenger* sea urchins, written by Harvard's Alexander Agassiz, one of many non-British scientists invited to work on the *Challenger* material by Charles Wyville Thomson.

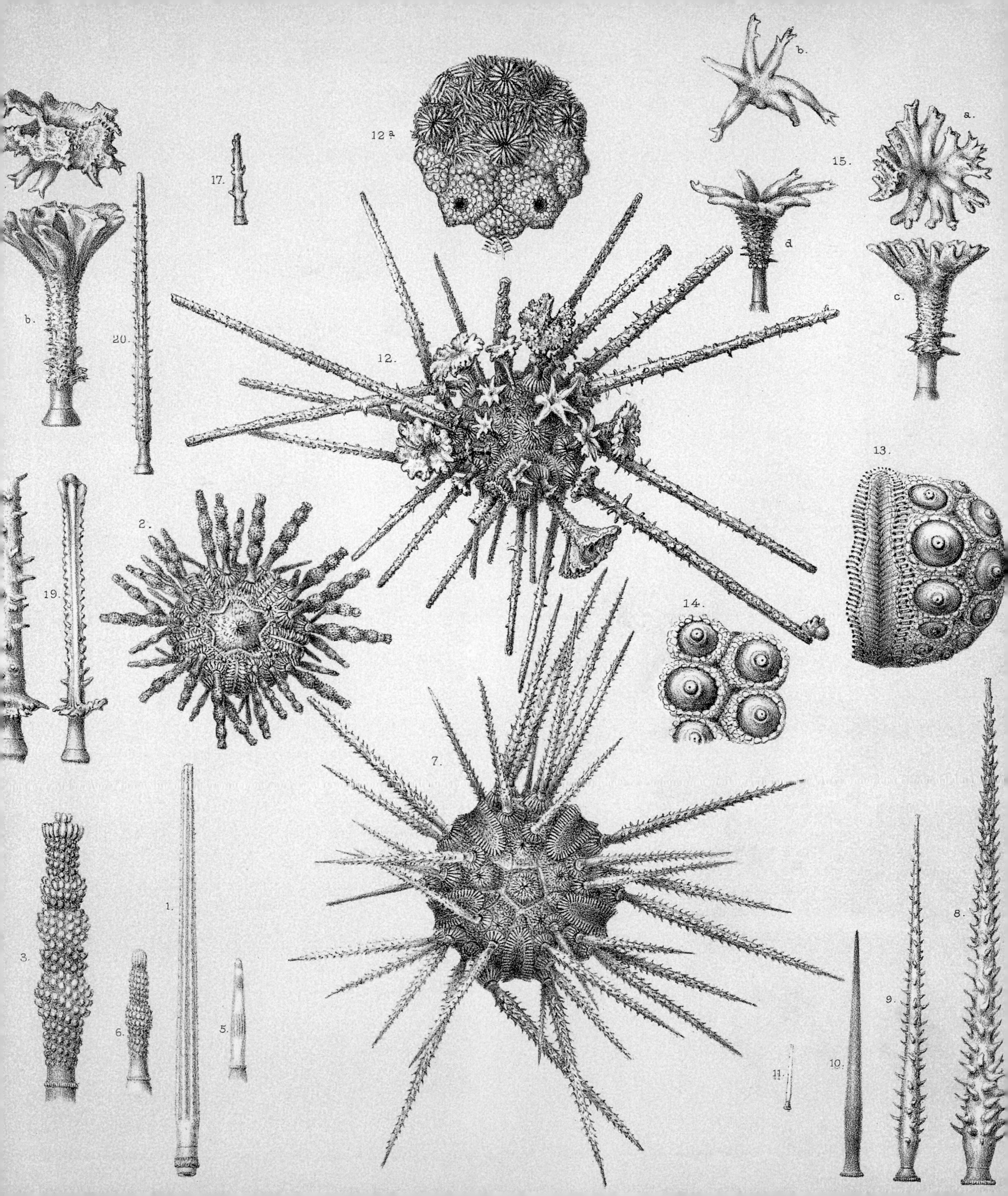
12 a.
b.
a.
15.
17.
d
c.
b.
20.
12.
13.
2.
19.
14.
7.
1.
3.
6.
5.
8.
9.
10.
11.

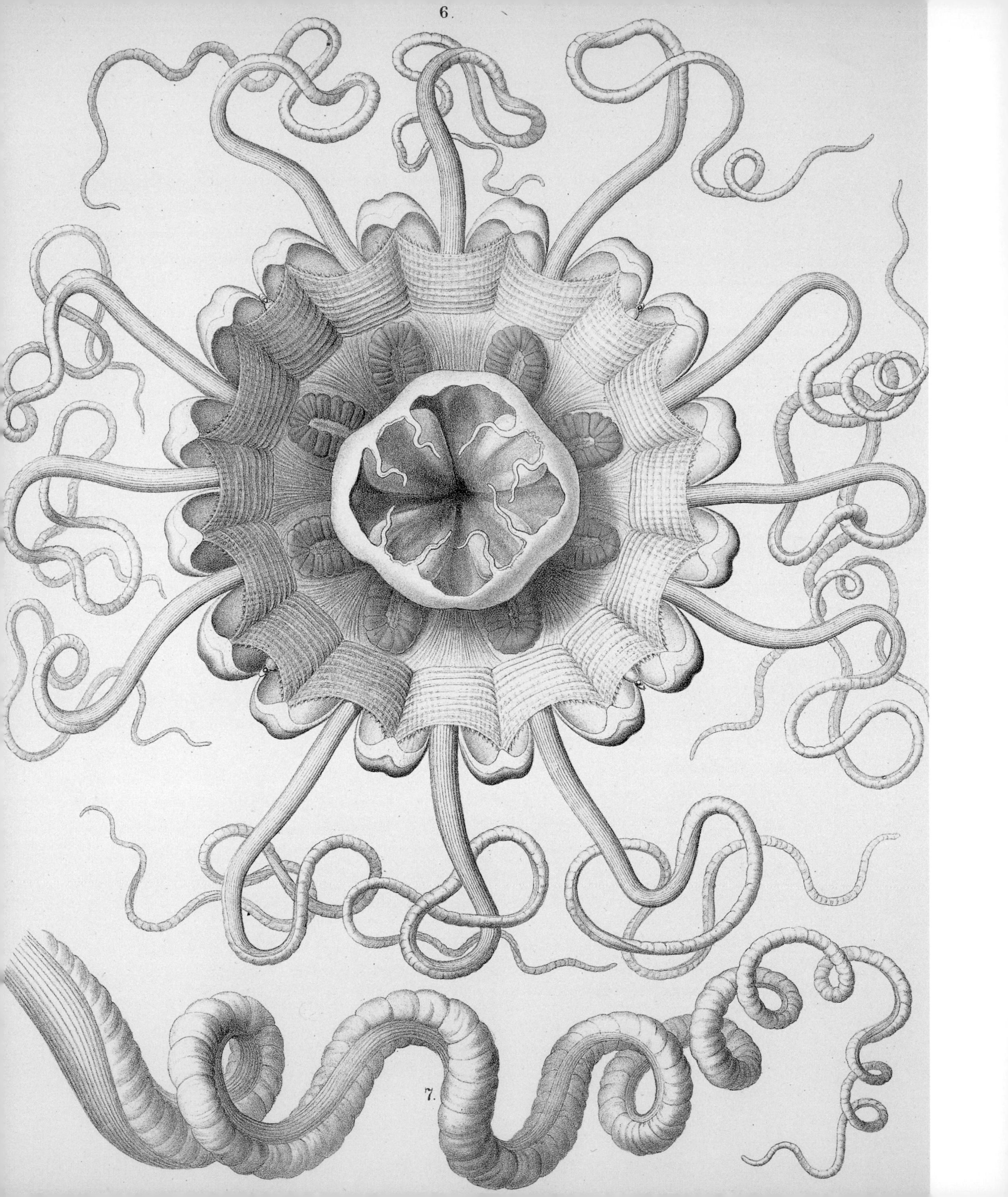
6.
7.

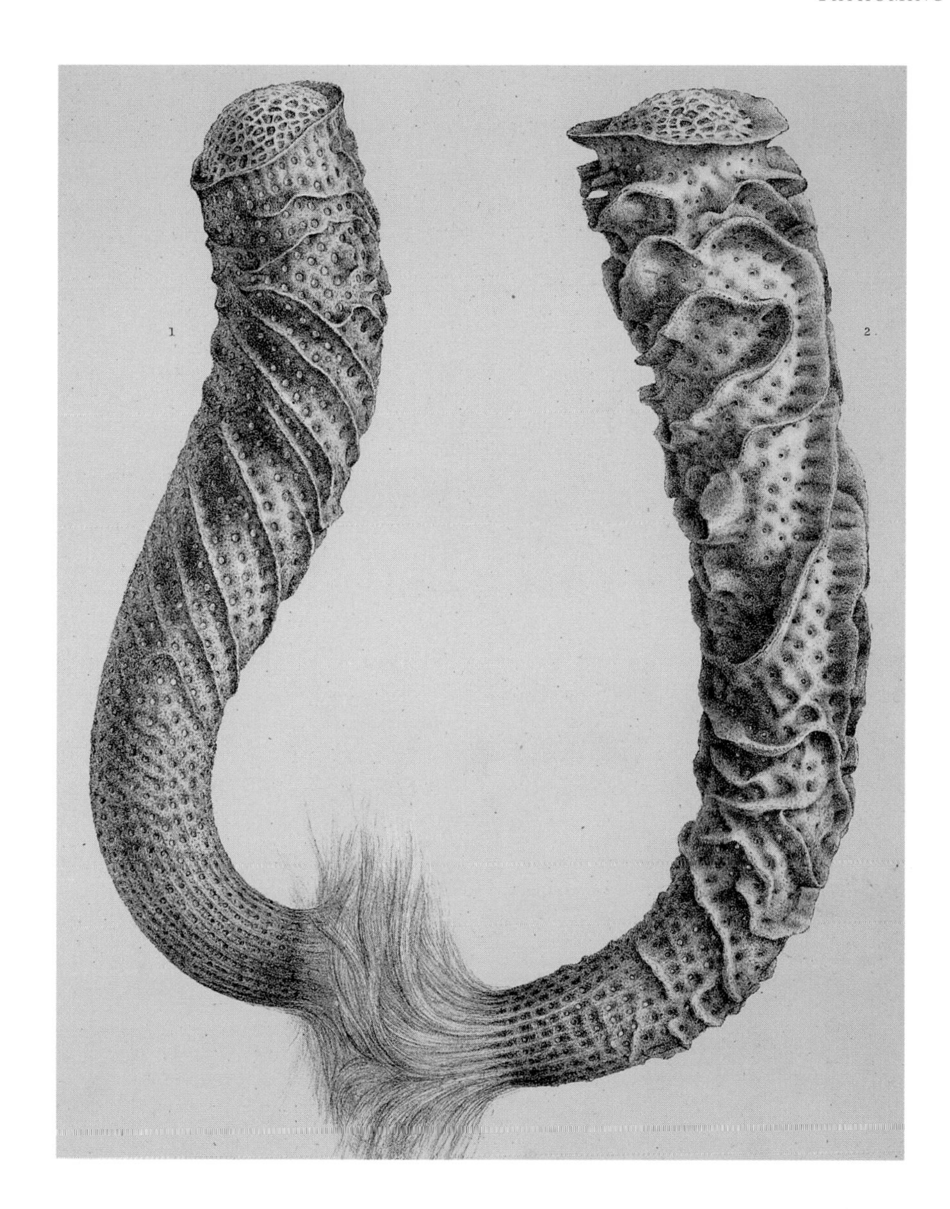

Many of the illustrations of animals collected from the *Challenger* are artistically accomplished. The deep mid-water jellyfish, *Periphylla mirabilis (opposite),* is depicted as you would see it in nature if you were beneath it and looking up towards the surface. Equally beautiful are two specimens of the Venus's Flower Basket sponge, *Euplectella aspergillum (above). Euplectella* sponges are also known as glass sponges because their elegant supporting skeletons, and their 'roots' which anchor them in the seabed, are made up of a complex system of needle-like spicules consisting of pure silica.

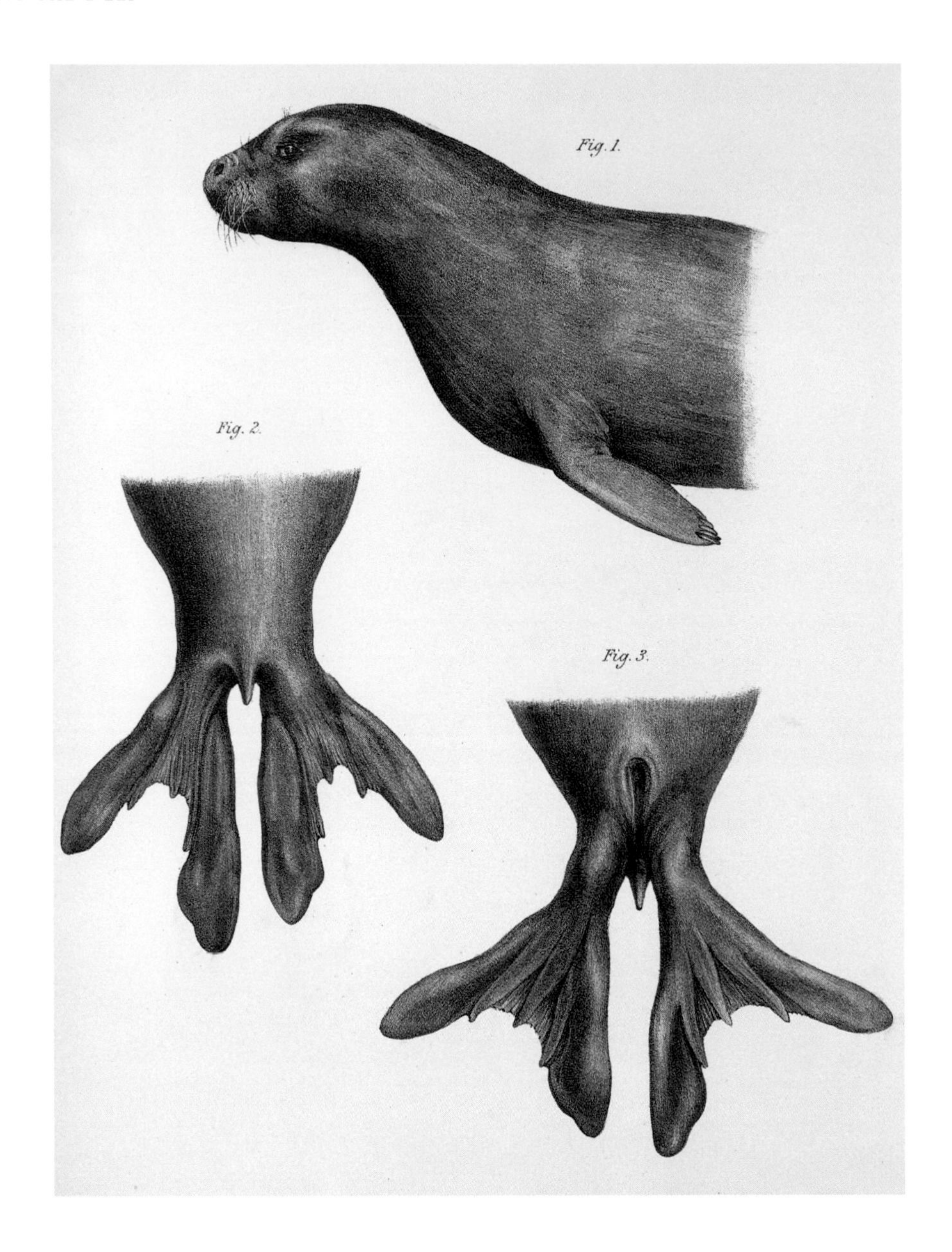

The front end and hind limbs of a female elephant seal, *Macrorhinus leoninus (above)* from Kerguelen in the southern Indian Ocean. Although the main objective of the *Challenger* voyage was deep-sea research, the participants also collected a range of ethnographic material, such as this human skull *(opposite)* from the Admiralty Islands in the south-western Pacific. Most of the specimens were artefacts such as tools, weapons and bowls, but some human remains were also brought home. Skulls, both human and animal, were commonly used to decorate the thatched roofs of the houses in the Admiralty Islands and the islanders were quite happy to sell them.

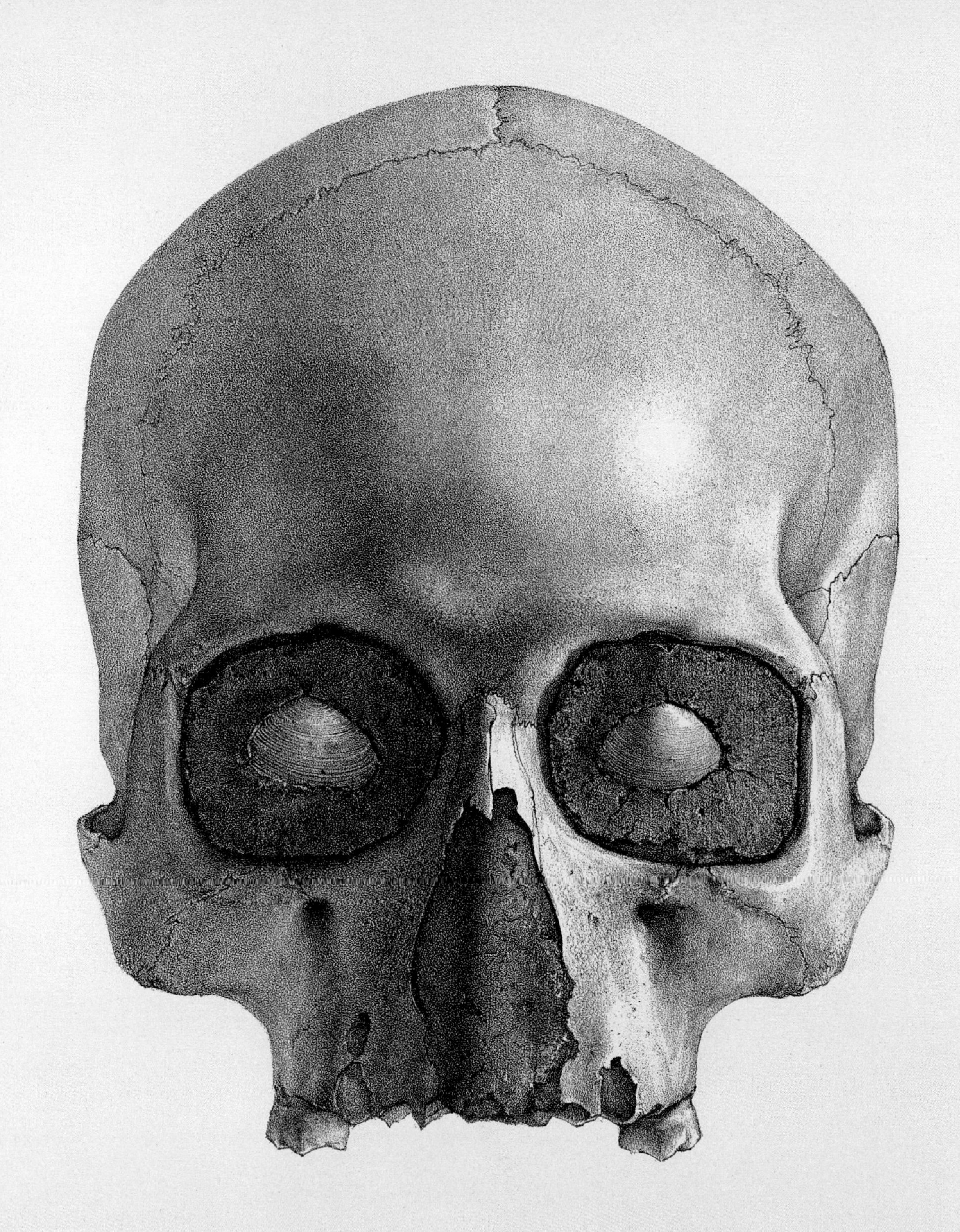

Epilogue

The journeys described in this volume cover an important and fascinating period in the history of science in general and of natural history in particular. It begins in the late 17th century, just as the post-medieval move towards rational or 'disinterested' enquiry into natural phenomena was gathering speed. Scientists, and especially natural historians, previously largely working in isolation, were becoming organised with the establishment of societies such as the Royal Society (1660) in England and the Académie des Sciences (1666) in France. They were communicating their results to one another in articles published in specialist journals, but it would be many years before science could provide a financially viable career for other than a very small number of university professors. As a result, for the next two centuries science was almost entirely the preserve of wealthy amateurs or the spare time pursuit of the educated middle classes, mainly clerics and doctors. The book ends in the late 19th century just as it was becoming possible to earn a living as a professional scientist, though rarely as financially rewarding as a successful career in law, finance, commerce, entertainment or, indeed, the arts. Yet throughout the intervening period one small, but notable group of professionals consistently benefited financially from science. These were the natural history artists, often paid but unsung heroes and heroines, whose legacy this book celebrates.

For the peripatetic botanists and zoologists of the period almost all needed the assistance of talented technicians to produce adequate images of the thousands of newly encountered plants and animals in their fresh state before they deteriorated into the dried and shrivelled specimens typical of most museum collections then as now. There was simply no other way to provide the objective accurate detail demanded by the new mood for realism characteristic of the 18th-century Enlightenment. And so the professional artist became a standard member of any scientific team sent out with European exploring expeditions, and the results of their labours ended up in private and public collections throughout the western world.

The last journey covered, the voyage of the *Challenger*, marks almost the end of this period, for the expedition not only had the services of an artist, but was also one of the first to use the relatively new technique of photography.

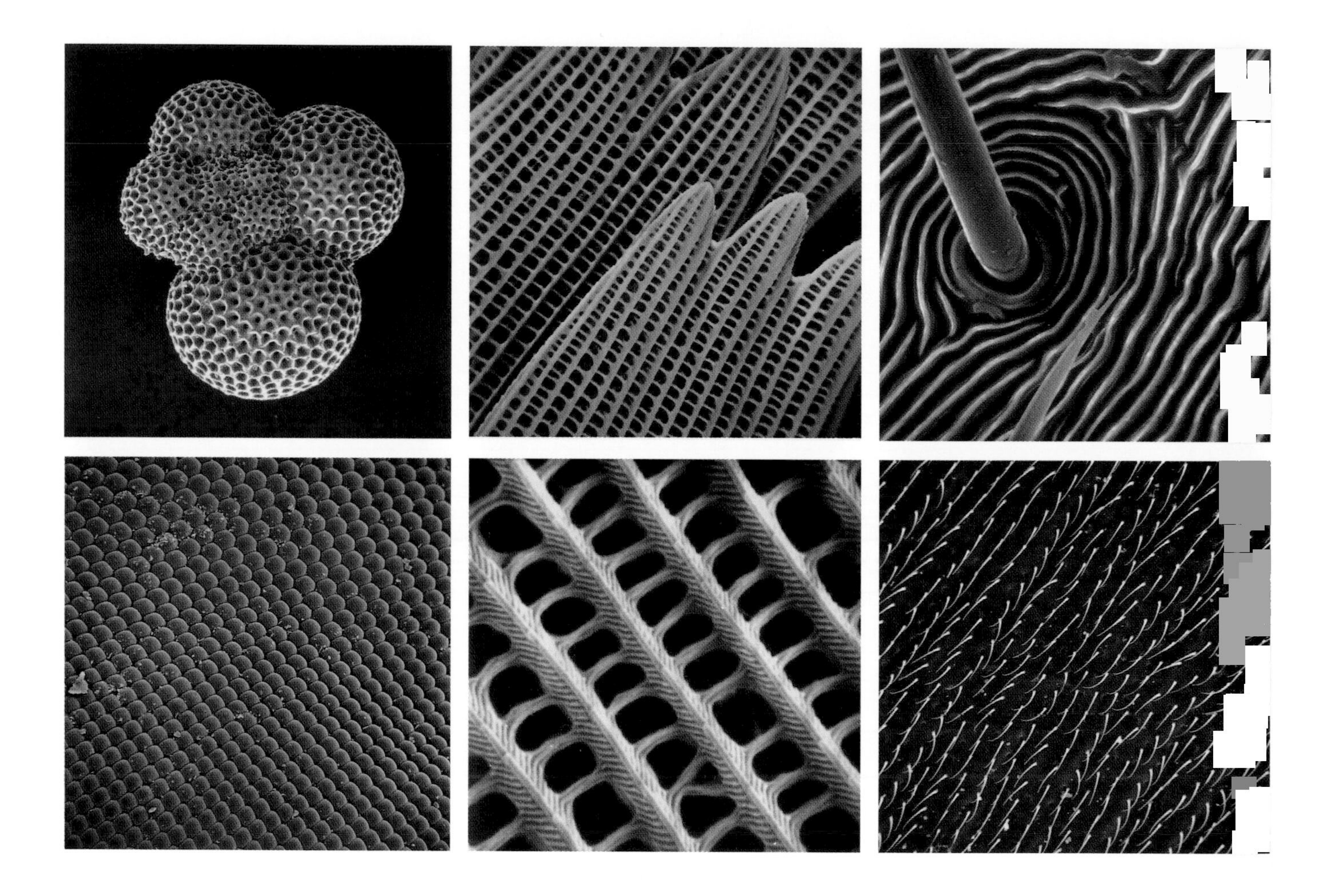

Modern scientific photography captures microscopic detail that naturalists and artists in previous centuries would have found difficult to imagine: the head of a small tortoiseshell butterfly *(previous page);* or *(opposite, clockwise from top left)* a foraminifer; the scale of a butterfly's wing; a spider cuticle; the surface of a blowfly wing; part of a butterfly wing scale; or the eye of a blowfly – magnified thousands of times by a scanning electron microscope.

Photography in the 1870s was still, of course, rather limited in its applicability. The cumbersome equipment involved, and the need to develop exposed plates more or less immediately, severely restricted its mobility. And the slow emulsions available meant that photography could be used only for immobile subjects, particularly views and rather stiff human portraits. But all this was to change within a few years. With the development of smaller and better equipment, faster films, colour and, of course, cinematography, the new technology was able to provide images that would have been quite impossible for the traditional artist. How would John Gould have reacted to an ultra high-speed photograph that 'stopped' the wings of one of his beloved hummingbirds moving at almost 100 beats a second – an image he could only guess at? Or Charles Wyville Thomson to photographs of animals on the deep-sea floor that had been thought to be totally lifeless only a few years before he sailed on the *Challenger*? Or any 18th- or 19th-century botanist to images from satellites showing the distribution of vegetational zones that they could never have imagined in their wildest dreams?

Photography did not make the artist redundant, however. Artists continued to accompany expeditions well into the present century and were employed to produce illustrations for scientific publications long after they were no longer routinely taken into the field. The Natural History Museum, for example, still had full-time artists on its staff in the 1960s. Their employment ended not because they were no longer needed, but because under restricted research funding their services were deemed to be marginally more expendable than those of the scientists. Notwithstanding the amazing developments in image technology, including holography, digital photography and the wonders of computer enhancement and virtual reality, no-one has been able to improve on the hand and the eye of a superb natural history artist when a particular morphological feature or the subtlest nuance of colour needs to be shown to maximum effect.

Illustration still has an important part to play in natural science. Claire Dalby's watercolour and pencil renderings of lichens *(above & opposite)* capture a sense of shape, texture and colour that photographs would be hard-pressed to match. Dalby is also able to create a perfect lichen specimen on paper, piecing together a number of fragments, where a photographer would rely on available, often damaged, specimens.

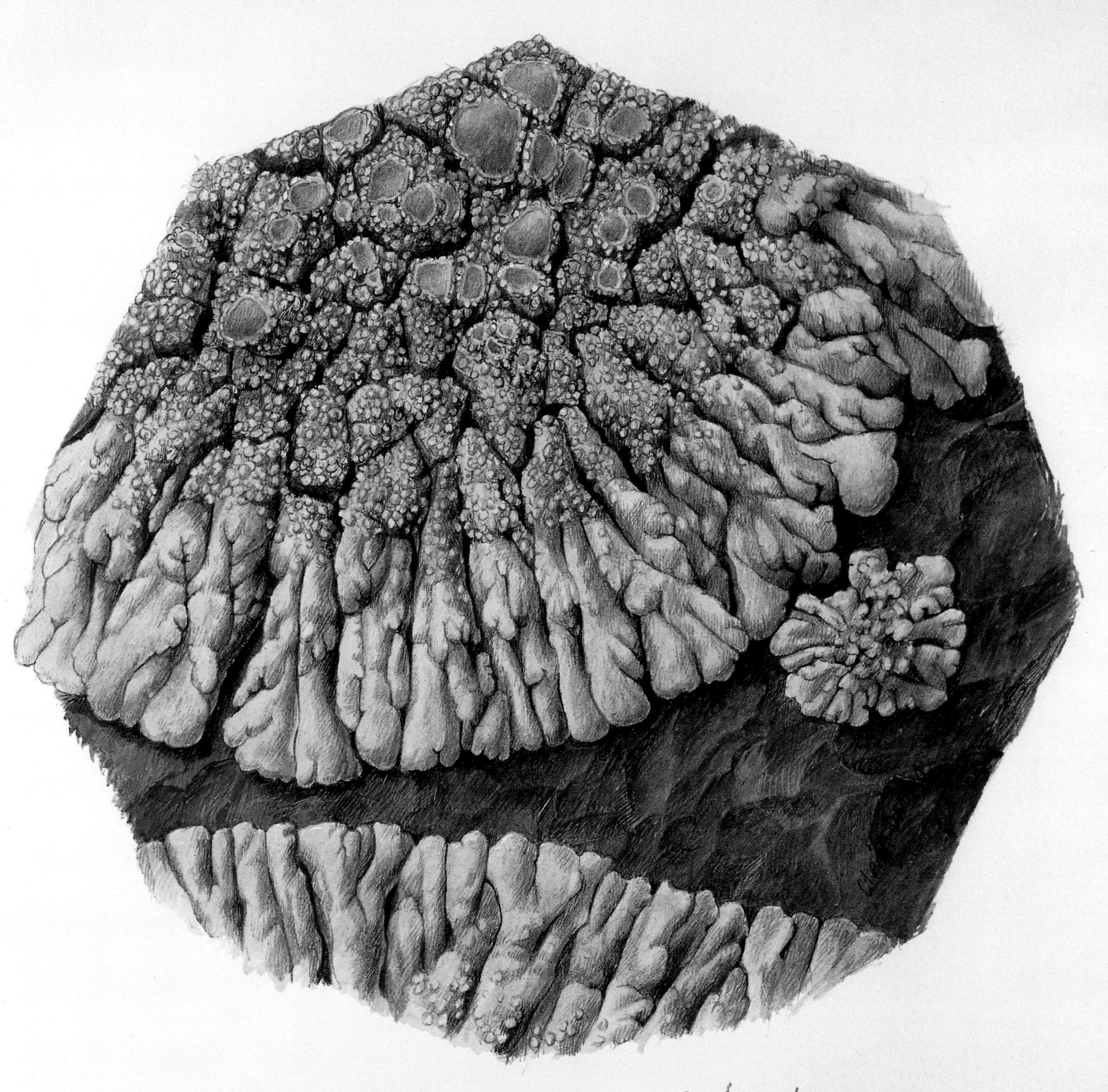

Caloplaca verruculifera
– Claire Dalby 1981

Selected Biographies

Joseph Banks

1743–1820 AMATEUR BOTANIST AND COLLECTOR

Banks was born in London, the son of a wealthy landowner. He attended the University of Oxford, where he employed a private tutor to teach him botany. He accompanied James Cook on the *Endeavour* to Australia and although he made few major voyages thereafter, he continued to accumulate books, manuscripts, drawings, paintings and specimens, particularly botanical ones. After Banks' death in 1820 his collections were entrusted to the British Museum.

John & William Bartram

1699–1777 & 1739–1823 LEADING AMERICAN HORTICULTURALISTS

John Bartram was born into a Quaker family in Pennsylvania. Although he had little formal education, he was keenly interested in botany and in 1729 founded his own botanical garden at Kingsessing near Philadelphia. He'd correspond with botanical enthusiasts in England and North America and introduced many North American plants to Europe. In 1743 he helped Benjamin Franklin establish the American Philosophical Society. William was born at Kingsessing in 1739. By the age of 15, William had been accompanying his father for several years on botanical field trips. In 1773, William embarked alone on a collecting expedition to the south-eastern states. William returned to Kingsessing in January 1777 six months after the American Declaration of Independence. John Bartram died eight months later.

Henry Walter Bates

1825–92 ENTOMOLOGIST AND COLLECTOR

Bates became fascinated by entomology as a teenager and published his first paper in the *Zoologist* at the age of 18. After he met Alfred Wallace in 1843, the two men travelled to the Amazon to study the local biology. Wallace stayed in South America for four years but Bates remained for 11 years, collecting more than 8,000 previously undescribed species, mostly insects. From his observations in the Amazon, Bates became convinced of the existence of biological evolution.

Ferdinand Bauer

1760–1826 ARTIST ON THE *INVESTIGATOR*

The Austrian-born son of an artist, Bauer inherited his father's talent and became particularly interested in botanical drawing. He worked as a botanical artist at the University of Vienna, before moving to England. He came to the attention of Joseph Banks who recommended him as artist on the *Investigator* voyage. On returning to England he worked for five years on his *Illustrationes Florae Novae Hollandiae*. In 1814 Bauer returned to Vienna where he continued to work as a botanical illustrator.

Robert Brown

1773–1858 BOTANIST ON THE *INVESTIGATOR*

Brown studied medicine in Edinburgh and joined the army as a surgeon's mate in 1795, but studied botany when off-duty. In 1798 he was offered the post of naturalist on the *Investigator* from 1801 to 1805, after Joseph Banks' first choice withdrew. Brown became 'Clerk Librarian and Housekeeper' of the Linnean Society from 1806 to 1822 and Banks' librarian and secretary from 1810. After inheriting Banks' collections, Brown gave them to the British Museum on condition that the herbarium formed the basis of a separate botanical department, of which he became the first Keeper.

James Cook

1728–79 ROYAL NAVAL OFFICER, RENOWNED EXPLORER, NAVIGATOR AND SURVEYOR

Cook was born at Marton in Yorkshire, the son of a farm labourer. He entered the Royal Navy in 1755, earning a reputation as an excellent surveyor. Cook was given command of three major voyages: the *Endeavour* to observe the transit of Venus from Tahiti and also to explore the southern Pacific; the *Resolution* and *Adventure* to fully investigate a possible southern continent; and the *Resolution* and *Discovery* to find a passage from the Pacific to the Atlantic across the north coast of America. Having failed to find the hoped-for northern passage, Cook turned back. When the ships reached Hawaii in January 1779, Cook was killed by islanders.

CHARLES DARWIN

1773–1858 NATURALIST ON HMS *BEAGLE*, AUTHOR OF *ON THE ORIGIN OF SPECIES*

Darwin was born in Shrewsbury, the son of a physician and grandson of physician, poet and philosopher Erasmus Darwin and of the potter Josiah Wedgwood. He studied medicine in Edinburgh but gave it up to study natural history. He was appointed naturalist and captain's companion on HMS *Beagle*, about to undertake a survey of the coast of South America under Robert FitzRoy. The ship was away from 1831 to 1836, during which time Darwin made extensive natural history and geological collections. These collections, and Darwin's simultaneous observations, formed the basis of his most famous work, *On the Origin of Species*, in which he set out his theory of evolution by natural selection.

PIETER CORNELIUS DE BEVERE

C.1733 ARTIST IN DUTCH CEYLON

Born in Colombo about 1733, the son of a junior officer in the Dutch East Indies Company, little is known of de Bevere's life apart from his employment by the colony's governor, Johan Gideon Loten in 1752 to illustrate the natural history of Ceylon, now Sri Lanka. When Loten moved to Batavia in 1757 he took de Bevere with him and continued to employ him until he returned to Europe. De Bevere seems to have died by 1781.

ROBERT FITZROY

1805–65 ROYAL NAVAL OFFICER, CAPTAIN OF HMS *BEAGLE*

The son of Lord Charles FitzRoy and descended from Charles II, Robert FitzRoy entered the navy in 1819 and was quickly promoted. He obtained his first command in 1828, surveying the coast of South America in HMS *Beagle*. He commanded the *Beagle* again from 1831 to 1836 to continue the South American survey. Charles Darwin accompanied the ship as naturalist and companion to FitzRoy, although FitzRoy's rather volatile character and deep religious convictions led to friction with Darwin over his ideas on evolution. After the *Beagle* returned, FitzRoy never went to sea again though he remained in the Navy for a further 14 years. He committed suicide in 1865.

MATTHEW FLINDERS

1774–1814 ROYAL NAVAL OFFICER, HYDROGRAPHER AND EXPLORER

Born in Lincolnshire, the son of a surgeon, Flinders entered the British navy in 1790. In 1795 he sailed in the *Reliance* to New South Wales. Over the next five years he surveyed south-eastern Australian waters, providing proof of the Bass Strait between the mainland and Tasmania. In 1801 Flinders was given command of the *Investigator* for an extended survey of the Australian coast. After completing his circumnavigation, Flinders embarked on an ill-fated trip home which took more than seven years. On his return to England in 1810 he began work on the official narrative of his voyage. He died in July 1814, the year his account was published.

JOHANN & GEORGE FORSTER

1729–98 & 1754–94 FATHER (NATURALIST) & SON (ARTIST) TEAM ON THE *RESOLUTION*

Johann was born at Dirschau near Danzig (now in Poland) in 1729 and studied theology, ancient and modern languages, medicine and natural history at the University of Halle. During 1766/7 he undertook a scientific and political survey of the new German colonies along the Volga for Catherine the Great accompanied by 11-year-old George. The family then moved to England,

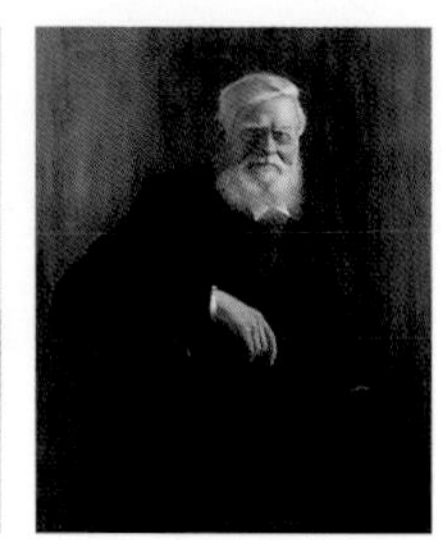

eventually settling in London where Johann wrote a number of papers, some illustrated by George. The pair became well known in the capital's scientific and antiquarian circles. They were appointed to the *Resolution* and did extremely valuable work during the voyage from 1772 to 1775. After falling out with the Admiralty over who should produce the report of the voyage, the Forsters returned to Germany.

John Gould

1804–81 ornithologist, artist and publisher of illustrated bird books

At 14, John Gould was apprenticed as a gardener at the Royal Gardens in Windsor where his father was foreman gardener. His training included taxidermy and this skill enabled him to obtain the post of curator and preserver for the newly established Zoological Society of London in 1827. Gould's interest in birds led to his appointment as the society's ornithological superintendent in 1833. In the succeeding decades he published many ornithological papers and was the first to recognise the significance of 'Darwin's finches' from the Galapagos Islands. Gould's parallel career was as a publisher of lavish books. By the time of his death he had published an impressive series of 41 folio volumes containing a total of 3,000 plates.

Paul Hermann

1646–95 botanist

Relatively little is known about Paul Hermann. In 1672, shortly after completing his medical studies in Holland, he became chief medical officer for the Dutch East Indies Company in Ceylon (now Sri Lanka). During his five years on the island Hermann made an extensive plant collection with accompanying illustrations. On his return to Holland he was appointed to the Chair of Botany at the University of Leiden in 1679. After his death in 1695 his herbarium was used by Carl Linnaeus as the basis of a number of descriptions of previously unknown species.

Carl Linnaeus

1707–78 instigator of the universally used system of botanical and zoological nomenclature

Linnaeus was born in the Swedish province of Småland. He studied medicine but his main love was botany. Linnaeus made a series of important plant-collecting journeys himself, but his many student 'disciples' travelled through the known world and sent him collections of plants which he used as the bases for a series of published works on regional flora. His work produced major advances in plant classification and in botanical and zoological nomenclature. His two-part or 'binomial' naming system, by which every plant and animal species received a unique Latin genus and species name, quickly became adopted by botanists and zoologists throughout the world and is the basis of the one used today.

Johan Gideon Loten

1710–89 amateur naturalist, Governor of Ceylon

Loten was born in Scadeshoeve, Holland, in 1710 and joined the staff of the Dutch East Indies Company in 1731. Having served in various capacities in Batavia, Samarang and Sulawesi, Loten was appointed Governor of Ceylon in 1752. During his five years on the island Loten employed local artist Pieter de Bevere to illustrate the island's flora and fauna. Loten returned to Holland, moved to England in 1759 and was elected a Fellow of the Royal Society in 1760. He returned to Holland in 1765.

Maria Sibylla Merian

1647–1717 naturalist and artist

Born in Frankfurt am Main, the daughter of a publisher and engraver, Merian worked as a flower painter and engraver but increasingly her work focused on entomology. In 1679 she published a book on the life cycle of European

Opposite (left to right): Charles Darwin, Henry Walter Bates in the Amazon, Captain James Cook, John Gould and Alfred Russel Wallace.

butterflies in which they were shown with their food plants, a new approach at the time. In 1699 she travelled to Surinam, illustrating butterflies and their host plants for her most famous work *Metamorphosis Insectorum Surinamensium*.

John Murray

1841–1914 NATURALIST ON THE *CHALLENGER*

Murray was born in Ontario, but in his teens went to live with his grandfather in Scotland. His medical studies and interest in marine life led to his appointment as a junior naturalist on the *Challenger.* A distinguished oceanographic career followed, for which he received a knighthood, fellowship of the Royal Society and honorary degrees from universities around the world.

Sydney Parkinson

1745–71 ARTIST ON THE *ENDEAVOUR*

Parkinson was born in Edinburgh, the son of a Quaker brewer. The Parkinsons moved to London where Sydney began exhibiting flower paintings, particularly on silk. His work impressed Joseph Banks who employed him as artist aboard the *Endeavour* from 1768 to 1771. During the voyage, he produced almost 1,000 plant drawings and 400 animal drawings, but he died on the journey home.

Hans Sloane

1660–1753 PHYSICIAN, BOTANIST AND COLLECTOR

Born at Killyleagh, Ireland in 1660, Sloane studied medicine and worked as a physician in London. Between 1687 and 1689 he lived in Jamaica as personal physician to the island's Governor. Sloane was already interested in natural history, particularly botany, and while in Jamaica amassed a large collection of plants and animals, which led to the publication of his *Natural History of Jamaica* between 1707 and 1725. During his lifetime he accumulated a vast collection of objects including natural history specimens, paintings, manuscripts, books and pamphlets, which formed the basis of the British Museum after his death in 1753.

Charles Wyville Thomson

1830–82 ZOOLOGIST AND SCIENTIFIC DIRECTOR OF THE *CHALLENGER*

Thomson was born near Linlithgow, Scotland, the son of a surgeon. He began medical studies in Edinburgh in 1845, but gave them up to study natural history. Thomson participated in a series of cruises in British waters which demonstrated the existence of life at great depths in the sea. The results were so encouraging that they led directly to the global oceanographic expedition in HMS *Challenger* from 1872 to 1876, with Thomson as chief scientist. He later oversaw publication of the expedition reports.

Alfred Russel Wallace

1823–1913 NATURALIST AND CO-PROPOSER OF THE MECHANISM OF NATURAL SELECTION

Through his studies of the distribution of animals on the Malay Archipelago from 1854 to 1862, English-born Wallace came to a very similar conclusion about their evolutionary origins as that being arrived at by Charles Darwin. Although not as well known as Darwin, in scientific circles he is acknowledged as co-proposer of the theory of evolution by natural selection.

John James Wild

1828–1900 ARTIST ON THE *CHALLENGER*

Wild was born in Switzerland and after studying in Zurich, Berne and Leipzig he moved to England to teach languages, eventually moving to Belfast where he met Charles Wyville Thomson, at the time Professor of Natural History at Queen's University. As a result of this friendship Wild was invited to be Thomson's secretary during the *Challenger* voyage and to act as the expedition's official artist. During the expedition Wild made many drawings and paintings, some of which were used in the official reports.

Bibliography

Chapter 1

Brooks, E. St. John, 1954. *Sir Hans Sloane. The Great Collector and his Circle.* London, the Batchworth Press, 234pp.

de Beer, G. R. 1953. *Sir Hans Sloane and the British Museum.* Oxford University Press, 192pp.

MacGregor, A. [Ed], 1994. *Sir Hans Sloane. Collector, Scientist, Antiquary Founding Father of the British Museum.* British Museum Press, 308pp, October 23, 1998.

Chapter 2

Maria Sibylla Merian, 1980. *Metamorphosis Insectorum Surinamensium (Amsterdam, 1705). Facsimile Edition,* Pion Ltd., London.

Rucker, E. & Stearn, W. T. 1982. *Maria Sibylla Merian in Surinam. Commentary to the Facsimile Edition of Metamorphosis Insectorum Surinamensium (Amsterdam, 1705).* Based on original watercolours in the Royal Collection, Windsor Castle. Pion, London.

Stearn, W. T. 1978. *The Wondrous Transformation of Caterpillars.* Scolar Press, 1978.

Wettengl, Kurt [Ed], 1997. *Maria Sibylla Merian 1647–1717, Artist and Naturalist.* Verlag Gerd Hatje, 275pp.

Chapter 3

Ferguson, D. *Joan Gideon Loten, F.R.S., the naturalist Governor of Ceylon (1752–57), and the Ceylonese Artist de Bevere,* J. Roy. Asiatic Soc. (Ceylon), 19: 217–268.

Trimen, H. 1887. *Hermann's Ceylon herbarium and Linnaeus's 'Flora Zeylanica'.* J. Linn. Soc. Botanical Series, 24: 129–155.

Blunt, W. 1984. *The Compleat Naturalist. A Life of Linnaeus.* William Collins Sons and Company Limited, 256pp.

Chapter 4

Bartram, W. 1791. *Travels Through North & South Carolina, Georgia, East & West Florida, the Cherokee Country, the extensive territories of the Muscogulges, or Creek Confederacy, and the Country of the Chactaws; containing an account of the soil and natural productions of those regions, together with observations on the manners of the Indians.* Philadelphia; James and Johnson, 522pp.

Ewan, Joseph, 1968. *William Bartram. Botanical and Zoological Drawings, 1756–1788.* American Philosophical Society, 180pp.

Fagin, N. Bryllion, 1933. *William Bartram, Interpreter of the American Landscape.* Baltimore, Johns Hopkins Press, 229pp.

Reveal, James L. 1992. *Gentle Conquest; the Botanical Discovery of North America with Illustrations from the Library of Congress.* Starwood Publishing Inc., 160pp.

Slaughter, Thomas P. 1996. *The Natures of John and William Bartram.* Knopf, 304pp.

Chapter 5

Beaglehole, J. G. 1974. *The Journals of Captain James Cook IV The Life of Captain James Cook.* The Hakluyt Society, London, 760pp.

Britten, J. 1900–1905. *Illustrations of Australian plants collected in 1770 during Captain Cook's voyage.*

Blunt, W. & Stearn, W. 1973. *Captain Cook's Florilegium,* Lion and Unicorn Press.

Carr, D. J. [Ed], 1983. *Sydney Parkinson; Artist of Cook's Endeavour Voyage.* London and Canberra: British Museum (Natural History), in association with Australian National University Press, XVI and 300pp.

Carter, H. B. 1988. *Sir Joseph Banks, 1743–1820.* British Museum (Natural History), London, 671pp.

Joppien, Rüdiger & Smith, Bernard, 1985. *The Art of Captain Cook's Voyages. Volume I. The Voyage of the Endeavour 1768–1771.* Oxford University Press in association with the Australian Academy of Humanities, Melbourne, 247pp.

Chapter 6

Beaglehole, J. G. 1974. *The Journals of Captain James Cook IV The*

Life of Captain James Cook. The Hakluyt Society, London, 760pp.
Carter, H. B. 1988. *Sir Joseph Banks, 1743–1820*. British Museum (Natural History), London, 671pp.
Joppien, Rüdiger & Smith, Bernard, 1985. *The Art of Captain Cook's Voyages, Vol 2, The Voyage of the Resolution and Adventure 1772–1775*. Melbourne, Oxford University Press in association with the Australian Academy of the Humanities, 274pp.
Whitehead, P. 1969. *Zoological specimens from Captain Cook's voyages*. Journal Society Bibliography of Natural History, 5 (3): 161–201.
Whitehead, P. 1978. *The Forster collection of zoological drawings in the British Museum (Natural History)*. Bulletin British Museum Natural History (historical series), 6 (2): 25–47.

Chapter 7

Brosse, Jacques, 1983. *Great Voyages of Exploration. The Golden Age of Discovery in the Pacific*. David Bateman Ltd., 232pp.
Edwards, P. I. 1976. *Robert Brown (1773–1858) and the natural history of Matthew Flinders's voyage in H.M.S. Investigator 1801–1805*. J. Soc. Biblphy nat. Hist. 7: 385–407.
Flinders, M. 1814. *A Voyage to Terra Australis ... in the years 1801, 1802, and 1803, in His Majesty's Ship the Investigator ... 2 vols. and atlas*. G. and W. Nicol, London.
Norst, Marlene J. 1989. *Ferdinand Bauer; The Australian Natural History Drawings*. British Museum (Natural History), 120pp.
Vallance, T. G. & Moore, D.T. 1982. *Geological aspects of the voyage of HMS Investigator in Australian waters, 1801–1805*. Bulletin of the British Museum (Natural History) Historical Series, 10 (1): 1–43.

Chapter 8

Desmond, A. & Moore, J. 1991. *Darwin*. Michael Joseph Ltd., 850pp.
Keynes, R. D. 1979. *The Beagle Record. Selections from the original pictorial records and written accounts of the voyage of H.M.S. Beagle*. Cambridge University Press, 409pp.
Moorehead, A. 1969. *Darwin and the Beagle*. Hamish Hamilton, London, 280pp.
Tree, I. 1991. *The Ruling Passion of John Gould. A biography of the bird man*. Barry and Jenkins Ltd., 250pp.

Chapter 9

Bate, H. W. 1863. *The Naturalist on the River Amazons*. London, John Murray.
Beddall, B. G. 1969. *Wallace and Bates in the Tropics*. London, Macmillan and Co., 241pp.
George, W. 1964. *Biologist Philosopher; a study of the life and writings of Alfred Russel Wallace*. Abelard-Schuman, London, 320pp.
Moon, H. P. 1976. *Henry Walter Bates F.R.S. 1825–1892. Explorer, Scientist and Darwinian*. Leicestershire Museums, Art Galleries and Records Services, 95pp.
Wallace, A. R. 1853. *A Narrative of Travels on the Amazon and Rio Negro, with an Account of the Native Tribes, and Observations on the Climate, Geology, and Natural History of the Amazon Valley*. London, Reeve and Co., 539pp.
Wallace, A. R. 1869. *The Malay Archipelago: the Land of the Orang-utan and the Bird of Paradise; a Narrative of Travel with Studies of Man and Nature*. London, Macmillan and Co., 653pp.

Chapter 10

Linklater, E. 1972. *The Voyage of the Challenger,* London, John Murray, 288pp.
Murray, J. 1895. *A summary of the scientific results ... Report on the Scientific Results of the Voyage of H.M.S. Challenger during ... 1873–76,* Summary, 1608pp, in 2 volumes, Stationery Office, London.
Wild, J. J. 1878. *At Anchor: A narrative of experiences afloat and ashore during the voyage of H.M.S. "Challenger" from 1872–76*. London and Belfast, Marcus Ward and Co., 198pp.

Index

Italics indicate captions; **bold** indicates main references; (b) indicates biographical notes.

Acknowledgments

The publishers gratefully thank all the staff at The Natural History Museum for their support and expertise, in particular the Director Neil Chalmers, Jane Hogg and Lynn Millhouse of the Publishing Division, Pat Hart of the Photographic Unit, Martin Pulsford and Lodvina Mascarenhas of the Picture Library, Malcolm Beasley, Neil Chambers, Ann Datta, Carol Gökçe, Julie Harvey, Ann Lum, Christopher Mills, John Thackray and the rest of the Library staff. The Museum's scientists Barry Clarke, Oliver Crimmen, Charlie Jarvis, Sandra Knapp, Dennis Adams, Colin McCarthy, Nigel Merrett, Alison Paul, Phil Rainbow, Frank Steinheimer, Roy Vickery and John Whittaker.

Special thanks also to: Dr David Bellamy, Eileen Brunton, Ute Heek, Tom Lamb and David Moore.

Picture Credits

All images sourced from the libraries of The Natural History Museum London, unless stated otherwise.

Please note that some of the images appearing in this book are details from the existing artworks.

Pages 4, 7, 9, 11 : Pencil drawings by Alfred Waterhouse.
Page 12 : Water colour of The Natural History Museum by Alfred Waterhouse, Victoria and Albert Museum, London.

Chapter 1: Pages 15, 24–49: ink drawings by various artists and original specimens collected by Hans Sloane from Volumes 1–7 of Sloane's Herbarium, Botany Dept. Page 17: artifact from the General Library. Page 18: from *A Voyage to Jamaica*, Botany Library. Pages: 21–23 printed plates from *A Voyage to Jamaica*, Vol. 2, Botany Library.

Chapter 2: Pages 57–59, 64–71: pencil drawings from Paul Hermann's Herbarium, Vol. 5, Botany Dept. Pages 60–63, 72–89: watercolours by Pieter de Bevere, General Library.

Chapter 3: Pages 91, 103, 108, 112: Facsimile 1981 edition of *Metamorphosis Insectorum Surinamensium* 1981, from original watercolours held in the Royal Collection, Windsor. Page 94: map library at British Library. Page 95: title pages 1705 and 1719 edition *Metamorphosis*. Pages 96–97, 100–119: Hand-coloured plates from 1705 and 1719 editions *Metamorphosis*. Pages 114–119: Plates appear only in 1719 edition. Pages 98–99: Hand-coloured plates from 1680 edition *Neues Blumenbuch*, Botany Library.

Chapter 4: Pages 121–141: drawings & watercolours by William Bartram, Botany Library.

Chapter 5: Page 143: ink and watercolour, attr. Port Jackson Painter, General Library, Page 145: ink and watercolour by Thomas Watling, General Library. Page 146: pages from Daniel Solander's handwritten journal *Plantae Novae Hollandiae*, Botany Library. Page 149: oil painting, self-portrait of Sydney Parkinson, Zoology Library. Page 150: title page, Daniel Solander's handwritten journal *Plantae Novae Hollandiae*, Botany Library. Engraving of Daniel Solander, 1784: drawn by James Sowerby, engraved by James Newton. Page 151: original specimen from Cryptogamic Herbarium. Pages 152–169: drawings and watercolours by Sydney Parkinson and related engravings, Botany Library.

Chapter 6: Pages 171, 174–197: watercolours by George Forster, Zoology Library. Page 173: catalogue handwritten by Johann Forster, General Library.

Chapter 7: Pages 199, 206–222, 224–229: watercolours by Ferdinand Bauer, Botany Library. Pages 200–201: plates by various artists from *A Voyage to Terra Australis* by Matthew Flinders, General Library. Page 202: ink plan of HMS *Investigator*, National Maritime Museum Greenwich. Page 205: Engraved print of Captain Flinders, National Maritime Museum Greenwich. Page 223: pencil drawing by Ferdinand Bauer, Naturhistorisches Museum Wien.

Chapter 8: Pages 231, 240–259: plates by various artists, including John and Elizabeth Gould, from *Zoology of the Voyage of the Beagle*. Vol. 1–3. Page 232: handwritten notes by Charles Darwin, General Library. Page 235: watercolour by Owen Stanley of HMS *Beagle*, 1841, National Maritime Museum Greenwich. Page 237: plate from 1870 edition of Darwin's *Journal of Researches*, General Library. Pages 238–239: plan of HMS *Beagle* from General Library.

Chapter 9: Pages 261, 263, 276–289: pencil and watercolour illustrations by Henry Walter Bates, Entomology Library. Page 262: Letter to John Gould from Alfred Russel Wallace, General Library. Page 265: Notebooks of Alfred Russel Wallace, General Library. Page 266: Photograph of Alfred Russel Wallace, General Library. Pages 268–271: Plates illustrated by John and Elizabeth Gould for John Gould's *Birds of New Guinea*, Vol. 1. Pages 272–275: pencil drawings by Alfred Russel Wallace, General Library.

Chapter 10: Pages 300–303: glass plate negatives by Caleb Newbold/Frederick Hodgeson/Jesse Lay, Palaeontology/Mineralogy Library. Pages 304–305: Drawings by J. J. Wild, National Maritime Museum Greenwich. Pages 291, 306, 308–319: plates by various artists from *The Challenger Reports*. Page 293: painting of 'HMS *Challenger* In The Ice' by W. F. Mitchell, National Maritime Museum Greenwich. Page 294: Royal Photographic Society. Page 297, 299: National Maritime Museum Greenwich. Page 298: artifacts, Palaeontology/ Mineralogy Library.

Epilogue: Page 321: computer-coloured photographic magnification, Picture Library. Page 322: scanning electron microscope images, Picture Library. Pages 324–325: watercolour and pencil, 1981, by Claire Dalby, Botany Library, reproduced courtesy of Claire Dalby. Page 328, from left: photograph, Picture Library; frontispiece of *The Naturalist on the River Amazons*, vol. 1, 1863 by Henry Walter Bates, Entomology Library; oil by Nathaniel Dance, National Maritime Museum Greenwich; crayon on paper drawing by Marion Walker, 1875, Zoology Library; photograph, Picture Library.